T0351317

Qualitative Theory of ODEs

An Introduction to
Dynamical Systems Theory

Qualitative Theory of ODEs

An Introduction to Dynamical Systems Theory

Henryk Żołądek
University of Warsaw, Poland

Raul Murillo
Complutense University of Madrid, Spain

NEW JERSEY • LONDON • SINGAPORE • BEIJING • SHANGHAI • HONG KONG • TAIPEI • CHENNAI • TOKYO

Published by

World Scientific Publishing Europe Ltd.

57 Shelton Street, Covent Garden, London WC2H 9HE

Head office: 5 Toh Tuck Link, Singapore 596224

USA office: 27 Warren Street, Suite 401-402, Hackensack, NJ 07601

Library of Congress Cataloging-in-Publication Data

Names: Żołądek, Henryk, 1953–   author. | Murillo, Raul, author.

Title: Qualitative theory of ODEs : an introduction to dynamical systems theory / Henryk Żołądek
    (University of Warsaw, Poland), Raul Murillo (Complutense University of Madrid, Spain).

Description: New Jersey : World Scientific, [2023] | Includes bibliographical references and index.

Identifiers: LCCN 2022021010 | ISBN 9781800612686 (hardcover) |
    ISBN 9781800612693 (ebook) | ISBN 9781800612709 (ebook other)

Subjects: LCSH: Differential equations--Qualitative theory. | Dynamics--Mathematical models.

Classification: LCC QA372 .Z643 2023 | DDC 515/.352--dc23/eng20220723

LC record available at https://lccn.loc.gov/2022021010

British Library Cataloguing-in-Publication Data

A catalogue record for this book is available from the British Library.

For any available supplementary material, please visit
https://www.worldscientific.com/worldscibooks/10.1142/Q0374#t=suppl

Desk Editors: Nimal Koliyat/Adam Binnie/Shi Ying Koe

Typeset by Stallion Press
Email: enquiries@stallionpress.com

Printed in Singapore

Preface

The Qualitative Theory of Ordinary Differential Equations occupies a rather special position both in Applied Mathematics and Theoretical Mathematics. On the one hand, it is a continuation of the standard course on Ordinary Differential Equations (ODEs). On the other hand, it is an introduction to Dynamical Systems, which is one of the main mathematical disciplines in recent decades. Moreover, it will turn out to be very useful for graduates when they encounter differential equations at their work; usually such equations are very complicated and cannot be solved by standard methods. Boasting aside, the first lecture on the Qualitative Theory of ODEs at the Faculty of Mathematics Informatics and Mechanics of Warsaw University was delivered by Henryk Żołądek in 1985; as you can see, the idea turned out to be successful.

The main idea of the qualitative analysis of differential equations is to be able to say something about the behavior of solutions of equations without solving them explicitly.

Therefore, in the first instance, such properties like the stability of solutions stand out. It is the stability with respect to changes in the initial conditions of the problem. Note that even with the numerical approach to differential equations, all calculations are subject to certain inevitable errors. Therefore, it is desirable that the asymptotic behavior of the solutions is insensitive to perturbations of the initial state. Chapter 1 of these lecture notes roughly focuses on this point.

Another important concept of this theory is structural stability. This is the stability of the entire system, e.g., the phase portrait, with respect to changes of parameters, which usually appear (and in large quantities) on the right side of the equations. In the absence of structural stability, we deal with bifurcations. Methods of the Qualitative Theory allow for precise and accurate investigation of such bifurcations. We describe them in Chapter 3.

In the case of two-dimensional autonomous systems, phase portraits are conceptually simple — they consist of singular points, their separatrices and limit cycles; one should add to this the resolution of singularities and behavior at infinity. It is worth mentioning that the problem of limit cycles for polynomial vector fields is the still unresolved Hilbert's 16th problem. These topics are discussed in Chapter 2.

Chapter 4 is dedicated to several issues in which a small parameter appears (in different contexts). It includes averaging (in different situations), the KAM theory and the theory of relaxation oscillations.

In multi-dimensional systems, new phenomena appear: transitivity, even distribution of trajectories, ergodicity, mixing and chaos. In Chapter 5, we discuss the ergodicity of translations on tori and dynamics of orientation-preserving diffeomorphisms of a circle.

The most elementary example of a chaotic system is the famous Smale horseshoe map, defined for a single transformation. In Chapter 5, we show how the Smale horseshoe appears in such elementary systems like a swing moved by periodic external force. We will also give other examples of chaotic behaviors, like Anosov systems and attractors.

In the Appendix (Chapter 6), the reader will find collected main facts from the course on ODEs. The last section, about Hamiltonian systems, is more advanced and rather minimal compared to the others; the reader must forgive us for this.

Each chapter contains a series of problems (with varying degrees of difficulty) and a self-respecting student should try to solve them.

This book is based on Raul Murillo's translation of Henryk Żołądek's lecture notes, which were in Polish and edited in the portal *Matematyka Stosowana (Applied Mathematics)* in the University of Warsaw.

At the end of the introduction, we would like to thank Zbigniew Peradzyński, who carefully read the Polish version of the manuscript and gave a list of comments and errors, and to Maciej Borodzik, whose useful remarks have helped to improve the text.

Henryk Żołądek was supported by the Polish NCN OPUS grant No. 2017/25/B/ST1/00931.

About the Authors

 **Henryk Żołądek** completed his education at the University of Warsaw in 1978. In 1983, he obtained his PhD degree from Moscow State University. He is working permanently at the University of Warsaw. He is also a coeditor (with G. Filipuk, A. Lastra, S. Michalik and Y. Takei) of the proceedings *Complex Differential and Difference Equations, de Gruyter Proceedings in Mathematics*, 2020. He is specializing in: Ordinary Differential Equations (limit cycles, normal forms), Mathematical Physics (classical and quantum mechanics) and Algebraic Geometry (algebraic curves). He is a member of the editorial board of the journals: *Topological Methods in Nonlinear Analysis, Qualitative Theory of Dynamical Systems, Journal of the Belarussian State University: Mathematics and Informatics*. He is also an author of nearly a hundred papers and the monograph *The Monodromy Group* (Birkhäuser, 2006). He has published his works in journals such as *Journal of Differential Equations, Journal of Statistical Physics, Israel Journal of Mathematics*, and *Lecture Notes in Mathematics*.

 **Raul Murillo** is a Mathematics and Computer Science graduate student. He received his BSc degree in 2019 from the Complutense University of Madrid. Here, from 2019 to 2021, he obtained an MSc degree in Computer Engineering, specializing in computer arithmetic. His main research interests include new computer arithmetic, approximate computing, high-performance computing, and deep neural networks. Throughout his academic career, he has undertaken stays in various European teaching and research centers, including the University of Warsaw and the Politecnico di Milano.

Contents

Preface   v

About the Authors   ix

1. Singular Points of Vector Fields   1

    1.1 Stability 2

    1.2 Hyperbolicity 17

    1.3 Exercises 30

2. Phase Portraits of Vector Fields   33

    2.1 Periodic solutions and focus quantities 35

    2.2 Poincaré–Bendixson criterion 48

    2.3 Index 55

    2.4 Dulac criterion 58

    2.5 Drawing phase portraits on the plane 66

        2.5.1 Singular points 66

        2.5.2 Closed phase curves 71

        2.5.3 Separatrices of singular points 71

        2.5.4 Behavior at infinity 72

        2.5.5 Orbital equivalence 73

    2.6 Exercises 75

3. Bifurcation Theory   79

    3.1 Versality 79

    3.2 Transversality 83

    3.3 Reductions 91

3.4   Codimension one bifurcations 100
3.5   Saddle-node bifurcation 101
3.6   Andronov–Hopf bifurcation 103
3.7   Codimension one non-local bifurcations 112
      3.7.1   Limit cycle bifurcations 112
      3.7.2   Separatrix connection bifurcations 115
3.8   Codimension two bifurcations 117
      3.8.1   Bogdanov–Takens bifurcation 117
      3.8.2   Saddle-node–Hopf bifurcation 120
      3.8.3   Hopf–Hopf bifurcation 123
      3.8.4   Limit cycle bifurcations with strong
              resonances 125
3.9   Exercises 128

4. Equations with a Small Parameter   133

4.1   Averaging 133
      4.1.1   Integrable Hamiltonian systems 133
      4.1.2   Averaging theorem 146
      4.1.3   Abelian integrals 148
      4.1.4   Subharmonic orbits 156
4.2   KAM theory 159
4.3   Partially integrable systems 167
4.4   Slow–fast systems 179
      4.4.1   Approaching the slow surface 181
      4.4.2   Movement along the slow surface 182
      4.4.3   Movement in the transition region 182
      4.4.4   Duck solutions 185
4.5   Exercises 187

5. Irregular Dynamics in Differential Equations   191

5.1   Translations on tori 192
5.2   Diffeomorphisms of a circle 194
5.3   Intersections of invariant manifolds 200
5.4   Smale horseshoe and symbolic dynamics 207
5.5   Ergodic aspects of dynamical systems 211
      5.5.1   Invariant measures 211
      5.5.2   Recurrence and ergodic theorem 214
      5.5.3   Mixing and entropy 216

5.6 Smooth dynamical systems 219
  5.6.1 Hyperbolicity 219
  5.6.2 Anosov diffeomorphisms 221
  5.6.3 Attractors 229
5.7 Exercises 230

6. Appendix: Basic Concepts and Theorems of the Theory of ODEs   **233**

6.1 Definitions 233
6.2 Theorems 236
6.3 Methods of solving 240
6.4 Linear systems and equations 244
6.5 Lagrangian and Hamiltonian systems 249
  6.5.1 Newtonian, Lagrangian and Hamiltonian mechanics 249
  6.5.2 Mechanics on Riemannian manifolds 252
  6.5.3 Canonical formalism 253
  6.5.4 Poisson brackets and integrability 254
  6.5.5 Symplectic and Poisson structures 257
6.6 Exercises 260

References   263

Index   265

Chapter 1

Singular Points of Vector Fields

Let us consider a non-autonomous system of differential equations (or a time-dependent vector field)

$$\dot{x} = v(t, x).$$

Here x belongs to a certain manifold M (the phase space) and t (time) to an interval $I \subset \mathbb{R}$. In this chapter, we can assume that M is an open subset of $\mathbb{R}^n$ and that the field v is of class C^r, $r \geq 2$. Thus, the assumptions of Appendix are fulfilled and the (local) solutions

$$x = \varphi(t; x_0, t_0)$$

to the above equation, satisfying the initial conditions $\varphi(t_0; x_0, t_0) = x_0$, exist.

Recall that a point x_* such that

$$v(t, x_*) = 0,$$

for every t, is called **equilibrium point**; other names in the literature are: **singular point** and **critical point** of the field (mainly in the case of autonomous fields). Of course, $\varphi(t) \equiv x_*$ is a solution to this problem.

The purpose of this chapter is to study properties of the solutions of the above system in a neighborhood of an equilibrium point.

1.1.   Stability

The simplest and desired property, from the point of view of applications, of the equilibrium point is its stability. In the following, we
give two mathematically rigorous definitions of stability.

Definition 1.1. An equilibrium point x_* of the equation $\dot{x} = v(t, x)$
is **stable in the Lyapunov sense** if for every $\varepsilon > 0$, there exists
a $\delta > 0$ such that, if $|x_0 - x_*| < \delta$, then for every $t > t_0$, we have
$|\varphi(t; x_0, t_0) - x_*| < \varepsilon$.
   An equilibrium point x_* is **asymptotically stable** if:

(i) it is stable in the Lyapunov sense, and
(ii) there exists $\varepsilon_0 > 0$ such that, if $|x_0 - x_*| < \varepsilon_0$, then
$\lim \varphi(t; x_0, t_0) = x_*$ as $t \to \infty$.

Example 1.2. For the *harmonic oscillator* $\ddot{x} = -\omega^2 x$, or

$$\dot{x} = y, \ \dot{y} = -\omega^2 x,$$

the solutions lie in the ellipses $\{(\omega x)^2 + y^2 = c\}$ (see Fig. 1.1). Therefore, if $0 < \omega \le 1$ then the product $\delta = \omega\varepsilon$ satisfies the definition
of stability in the Lyapunov sense. Since generic solutions do not
approach the equilibrium point $x = y = 0$, it is not asymptotically
stable.

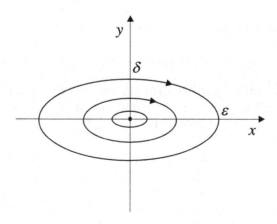

Fig. 1.1.   Harmonic oscillator.

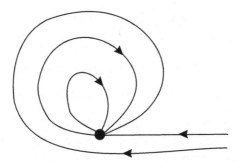

Fig. 1.2.   Not Lyapunov stable.

Example 1.3. Figure 1.2 shows the phase portrait of a vector field which has the property that each solution approaches the equilibrium point, i.e., the second of the conditions for asymptotic stability is fulfilled. However, this equilibrium point is not stable in the Lyapunov sense, since the trajectories starting from the bottom and arbitrarily close to the equilibrium point pass away of a fixed neighborhood of this point.

It turns out that the appropriate autonomous vector field can be specified by an explicit formula. Namely, it has the form

$$\dot{x} = y, \ \dot{y} = -2x^3 - 4xy - y(x^2 + y^2)^2 \tag{1.1}$$

(see Example 2.44 in the next chapter).

The main result about the stability of equilibrium points is due to A. Lyapunov. It concerns the equilibrium point $x = 0$ for a germ[a] of an autonomous vector field in $(\mathbb{R}^n, 0)$ of the form

$$v(x) = Ax + O(|x|^2), \tag{1.2}$$

where $A = \frac{\partial v}{\partial x}(0)$ is the linearization matrix of the field at $x = 0$.

---

[a]By a germ of vector field $v(x)$ (or of function $f(x)$ or of differential form $w(x)$ or of mapping) at a point $x_0 \in \mathbb{R}^n$ we mean a vector field (or a function or a differential form or a mapping) defined on some neighborhood U of point x_0. Two germs, one defined on a neighborhood U and the other on U', are equivalent if they are compatible in certain neighborhood $V \subset U \cap U'$. The notation $f : (\mathbb{R}^n, x_0) \to \mathbb{R}$ is used to denote the germ of a function in x_0; analogous notations are for vector fields, differential forms, mappings, etc.

Theorem 1.4 (Lyapunov). *If the linearization matrix A has the property that the real parts of all its eigenvalues are negative,*

$$\operatorname{Re}\lambda_j < 0, \quad j = 1, \ldots, n, \tag{1.3}$$

then the equilibrium point $x = 0$ is asymptotically stable.

Before we begin the very proof this theorem, we introduce the concept of Lyapunov function, which turns out to be useful for showing asymptotic stability even without assumption (1.3).

Definition 1.5. The **Lyapunov function** for the equilibrium point $x = 0$ of the autonomous vector field $v(x)$ is a function

$$L : U \to \mathbb{R}$$

from a neighborhood U of the point $x = 0$, which satisfies the following two properties:

(1) $L(x) \geq 0$ and $L(x) = 0$ if and only if $x = 0$;
(2) $\dot{L}(x) = \langle dL(x), v(x) \rangle < 0$ for $x \neq 0$.

Proposition 1.6. *If there exists a Lyapunov function (for the equilibrium point $x = 0$ of the field $v(x)$), then this point is asymptotically stable.*

Proof. Property (1) from the definition of the Lyapunov function says that the sets $\{L(x) \leq c\}$, $c > 0$, are bounded and tend to the point $x = 0$ at $c \to 0$.

Property (2) means that, if $x = \varphi(t)$ is the solution of equation $\dot{x} = v(x)$, then

$$\frac{d}{dt}(L \circ \varphi(t)) = \frac{\partial L}{\partial x}(\varphi(t)) \cdot \dot{\varphi}(t) = (\nabla L(x), v(x)) = \langle dL(x), v(x) \rangle < 0.$$

It implies that the Lyapunov function decreases along the solutions of the differential equation (see Fig. 1.3).

So, the solutions starting from the boundary $\{L(x) = c\}$ of the set $\{L(x) \leq c\}$ "enter" into the interior of this set. Since these trajectories remain in the sets $\{L \leq c\}$, the condition of stability in the Lyapunov sense is satisfied.

On the other hand, a solution cannot accumulate at a fixed level $\{L = c_0\}$, $c_0 > 0$, because of Flow Box Theorem (Theorem 6.10

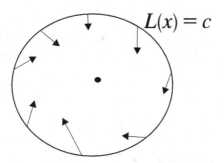

Fig. 1.3. Lyapunov function.

from Appendix) and of the compactness of this level. So, it must tend to the point $x = 0$ as $t \to \infty$ and this means the asymptotic stability. □

Now, for the proof of Lyapunov's theorem, it is sufficient to construct the Lyapunov functions. For this purpose, we will slightly improve the matrix A. Firstly, we assume that it is in the Jordan form. So, we have blocks

$$\begin{pmatrix} \lambda_j & 1 & 0 & 0 & \dots \\ 0 & \lambda_j & 1 & 0 & \dots \\ 0 & 0 & \lambda_j & 1 & \dots \\ \dots & \dots & \dots & \dots & \dots \\ 0 & 0 & 0 & 0 & \lambda_j \end{pmatrix}, \quad \begin{pmatrix} \alpha_j & -\beta_j & 1 & 0 & \dots \\ \beta_j & \alpha_j & 0 & 1 & \dots \\ 0 & 0 & \alpha_j & -\beta_j & \dots \\ 0 & 0 & \beta_j & \alpha_j & \dots \\ \dots & \dots & \dots & \dots & \dots \end{pmatrix},$$

corresponding to real (λ_j, $j = 1, \dots, r$) and complex ($\lambda_j = \bar{\lambda}_{j+1} = \alpha_j + i\beta_j$, $j = r+1, r+3, \dots, n-1$) eigenvalues.

It turns out that the 1's over the diagonal can be replaced with small ε's. Indeed, if we have a Jordan k–dimensional block with a real eigenvalue of λ then, in the standard basis (e_j), we have

$$Ae_1 = \lambda e_1, \ Ae_2 = \lambda e_2 + e_1, \dots, \ Ae_k = \lambda e_k + e_{k-1}.$$

So, for the basis (f_j), such that

$$f_k = e_k, \ f_{k-1} = \varepsilon^{-1} e_{k-1}, \dots, \ f_1 = \varepsilon^{1-k} e_1,$$

we will have $Af_1 = f_1$ and $Af_j = \lambda f_j + \varepsilon f_{j-1}$ $(j > 1)$. We use an analogous change when we have a Jordan block with complex eigenvalues (Exercise 1.28). So, we have the following lemma.

Lemma 1.7. *There exists a linear coordinate system such that the matrix A takes the form*

$$A = A_0 + \varepsilon A_1,$$

where A_0 is block-diagonal with $\lambda_j \in \mathbb{R}$ or $\begin{pmatrix} \alpha_j & -\beta_j \\ \beta_j & \alpha_j \end{pmatrix}$ on diagonals, and the matrix A_1 is bounded, $\|A_1\| < C_1$.

The following lemma completes the proof of Theorem 1.4.

Lemma 1.8. *Let (x_i) be the coordinate system from the thesis of Lemma 1.7. Then the function*

$$L(x) = \sum x_i^2 = (x, x) = |x|^2$$

is a Lyapunov function for this equilibrium point in a suitably small neighborhood of the point $x = 0$.

Proof. Of course, we only need to check property (2) from the definition of Lyapunov function. We have

$$\dot{L} = (\nabla L, A_0 x) + \varepsilon(\nabla L, A_1 x) + (\nabla L, v - Ax),$$

where $\nabla L = 2x$. The first term in the right-hand side of this equality is (as one easily checks)

$$(\nabla L, A_0 x) = 2 \sum_{j=1}^{r} \lambda_j x_j^2 + 2 \sum \alpha_j (x_j^2 + x_{j+1}^2), \qquad (1.4)$$

where in the second sum we take $j = r + 1, r + 3, \ldots, n - 1$. Next, with the bound for A_1, we get

$$|(\nabla L, A_1 x)| \leq 2C_1 |x|^2.$$

Since the nonlinear terms $v(x) - Ax$ are of order $O(|x|^2)$, we have

$$|(\nabla L, v - Ax)| \leq 2C_2 |x|^3 \leq 2C_2 \varepsilon |x|^2$$

for some constant C_2 and sufficiently small $|x|$.

Condition (1.3) from the assumption of Lyapunov's theorem means that in Eq. (1.4) we have

$$\lambda_j, \alpha_k < -\Lambda < 0$$

for some Λ; so, we have $(\nabla L, A_0 x) < -2\Lambda |x|^2$. The other two terms in $\dot{L}$ are estimated by $2(C_1 + C_2)\varepsilon |x|^2$. This shows that $\dot{L} < 0$ for $x \neq 0$ and small ε, which concludes the proof of the lemma and of the Lyapunov's theorem. □

There is a theorem inverse to the Lyapunov's theorem. Presumably Lyapunov was aware about it, but in Russian literature (e.g., in [1,2]) it is attributed to V. Chetaev.

Theorem 1.9 (Chetaev). *If the linearization matrix A of the vector field (1.2) has an eigenvalue with positive real part, then the equilibrium point $x = 0$ is not stable (neither in Lyapunov sense nor asymptotically).*

Proof. Let $\operatorname{Re} \lambda_1, \ldots, \operatorname{Re} \lambda_k > 0$ and $\operatorname{Re} \lambda_{k+1}, \ldots, \operatorname{Re} \lambda_{k+l} \leq 0$, $n = k + l$. We can assume that

$$A = A_1 \oplus A_2$$

in the splitting $\mathbb{R}^n = \mathbb{R}^k \oplus \mathbb{R}^l$, where the matrix A_1 has the eigenvalues $\lambda_1, \ldots, \lambda_k$ and the matrix A_2 has the eigenvalues $\lambda_{k+1}, \ldots, \lambda_{k+l}$. In addition, we can assume that the matrices A_1 and A_2 are like in Lemma 1.7. Let us assume that $x = (x_1, x_2)$ in the above splitting of $\mathbb{R}^n$ and put $|x| = |x_1| + |x_2|$.

Let us define a cone V using the inequalities

$$|x_2| \leq \alpha |x_1|, \quad |x_1| \leq \beta,$$

where constants α and β will be defined during further stages of the proof. Note that the boundary ∂V of cone V consists of two parts: $\partial_1 V = \{|x_2| = \alpha |x_1|\}$ and $\partial_2 V = \{|x_1| = \beta\}$. Let us also define the *Chetaev function* as

$$C(x) = |x_1|.$$

It turns out that, with appropriately chosen α and β, we have the following properties:

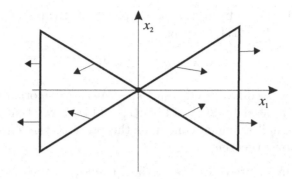

Fig. 1.4.   Chetaev function.

(1) the vector field is directed to the interior of V on the part $\partial_1 V$,
(2) $\dot{C}(x) > 0$ for $x \in V \setminus 0$.

Of course, the thesis of the theorem follows; any trajectory starting arbitrarily close to $x = 0$ in V goes out of V through the part $\partial_2 V$ of the boundary (see Fig. 1.4).

To prove the above properties we will use the following inequalities (which are consequences of the assumptions made):

$$\frac{\mathrm{d}}{\mathrm{d}t}|x_1| > M|x_1| - \varepsilon|x|, \quad \frac{\mathrm{d}}{\mathrm{d}t}|x_2| < \varepsilon|x|,$$

(for $M = \min \operatorname{Re}\lambda_j : 1 \leq j \leq k$) and small ε, with the condition that $|x| < \beta$ (β sufficiently small). As usual, $\mathrm{d}/\mathrm{d}t$ denotes the derivative along the trajectory $x(t)$ of the vector field.

Condition (1) means that $\frac{\mathrm{d}}{\mathrm{d}t}(|x_2| - \alpha|x_1|)|_{\partial_1 V} < 0$. But for $|x_1| = \alpha|x_2|$ we have $|x| = (1 + \alpha)|x_1|$ and

$$\frac{\mathrm{d}}{\mathrm{d}t}(|x_2| - \alpha|x_1|) \leq \varepsilon|x| - \alpha M|x_1| + \alpha\varepsilon|x|$$

$$= -\left(\alpha M - (\alpha + 1)^2\varepsilon\right)|x_1| < 0,$$

if ε is small. On the other hand, for $|x_2| \leq \alpha|x_1|$, we have

$$\frac{\mathrm{d}}{\mathrm{d}t}|x_1| > (M - (\alpha + 1)\varepsilon)|x_1| > 0.$$

$\square$

In connection with the above theorems, a natural practical question arises:

How to check whether all the eigenvalues of a given matrix have negative real parts?

Of course, this question boils down to the question of the real parts of the roots of the characteristic polynomial of this matrix.

So, assume that we have a polynomial[b]

$$P(\lambda) = a_0\lambda^n + a_1\lambda^{n-1} + \cdots + a_n, \quad a_0 > 0, \ a_j \in \mathbb{R}. \qquad (1.5)$$

Definition 1.10. We say that the polynomial $P(\lambda)$ is **stable** if all its zeros λ_j have negative real parts, $\operatorname{Re}\lambda_j < 0$.

We ask for the necessary and sufficient conditions for the polynomial (1.5) to be stable. It turns out that this problem was tested in the 19th century and has a complete solution.

To look at this issue, let us note the following simple necessary condition.

Lemma 1.11. *If a polynomial of form* (1.5) *is stable, then* $a_j > 0$ *for all* j.

Proof. Let us look at the factors in the polynomial

$$P(\lambda) = a_0 \prod(\lambda - \lambda_j) \prod(\lambda^2 - 2\alpha_j\lambda + (\alpha_j^2 + \beta_j^2)),$$

where the first product is related to the real roots $\lambda_j < 0$ and the second product is related to the unreal roots $\lambda_j = \alpha_j \pm i\beta_j$, $\alpha_j < 0$, $\beta_j \neq 0$. Since each factor has positive coefficients, then the whole polynomial must also have positive coefficients. □

Remark 1.12. If the degree $n \leq 2$, then the condition $a_j > 0$, $j = 0, 1, 2$, is also a sufficient.

---

[b]The polynomial of $\det(A - \lambda)$ has the coefficient $a_0 = (-1)^n$. Here we take $a_0 > 0$ to simplify the formulated results.

Define the following $n \times n$ dimension matrix:

$$M = \begin{pmatrix} a_1 & a_0 & 0 & 0 & \ldots & 0 & 0 \\ a_3 & a_2 & a_1 & a_0 & \ldots & 0 & 0 \\ a_5 & a_4 & a_3 & a_2 & \ldots & 0 & 0 \\ \vdots & \vdots & \vdots & \vdots & \ddots & \vdots & \vdots \\ 0 & 0 & 0 & 0 & \ldots & a_{n-1} & a_{n-2} \\ 0 & 0 & 0 & 0 & \ldots & 0 & a_n \end{pmatrix}, \qquad (1.6)$$

such that on the diagonal stand consecutively the numbers $a_1, a_2, \ldots, a_n$. This is called the **Hurwitz matrix.**

Theorem 1.13 (Routh–Hurwitz conditions). *A necessary and sufficient condition for the stability of the polynomial* (1.5) *is:*

(i) $a_j > 0$ *for all* j;
(ii) *the principal minors* Δ_j *of the matrix* (1.6) *are positive.*

Moreover, if any of the above inequalities is reversed, i.e., $a_j < 0$ *for some* j *or* $\Delta_k < 0$ *for some* k, *then some of the zeroes of the polynomial* $P(\lambda)$ *has positive real part.*

Example 1.14. For $n = 1$, the matrix (1.6) has the form $M = (a_1)$, so $\Delta_1 = a_1$.

For $n = 2$, that is for the matrix $M = \begin{pmatrix} a_1 & a_0 \\ 0 & a_2 \end{pmatrix}$, we have $\Delta_1 = a_1$ and $\Delta_2 = a_2 a_1$; therefore, we are under the conditions of Remark 1.12.

For $n = 3$, we have the matrix

$$\begin{pmatrix} a_1 & a_0 & 0 \\ a_3 & a_2 & a_1 \\ 0 & 0 & a_3 \end{pmatrix}.$$

The Routh–Hurwitz conditions take the form: $\Delta_1 = a_1 > 0$ (nothing new),

$$\Delta_2 = a_1 a_2 - a_0 a_3 > 0 \qquad (1.7)$$

and $\Delta_3 = a_3 \Delta_2$ (also nothing new).

Remark 1.15. It can be shown that the condition $\Delta_j > 0$ for all j can be replaced by the following Liénard–Shapiro conditions:

$$\Delta_2 > 0, \ \Delta_4 > 0, \ \Delta_6 > 0, \ldots$$

(see also the proof follows).

Proof of Theorem 1.13. The idea of the proof is quite simple. The conditions $\operatorname{Re} \lambda_j < 0$ (and $a_0 > 0$) define a certain subset U in the space $\mathbb{R}^{n+1} = \{a\}$ of the coefficients a_j. The set U is semi-algebraic and its boundary ∂U consists of smooth "strata". We are interested in the equations that define these strata. If you have $a \in \partial U$, then there are two possibilities:

(a) a root of the equation $P(\lambda) = 0$ is zero,
(b) a pair of complex conjugate roots lies on the imaginary axis.

Case (a) means that $P(0) = 0$, i.e., $a_n = 0$; this is quite simple.
Let us consider the situation with a pair $\lambda_{j,j+1} = \pm i\beta$ of imaginary roots. We have then

$$P(\lambda) = (\lambda^2 + \beta^2) Q(\lambda) \qquad (1.8)$$

for a certain polynomial

$$Q = b_0 \lambda^{n-2} + b_1 \lambda^{n-3} + \cdots + b_{n-2},$$

which we can assume that is stable. In addition, from the inductive assumption (with respect to n), we can assume that $b_j > 0$ and the corresponding minors $\Delta_j = \Delta_j(Q) > 0$.

We have the following relations:
$$a_0 = b_0, \; a_1 = b_1, \; a_2 = b_2 + \beta^2 b_0, \; a_3 = b_3 + \beta^2 b_1, \ldots$$

This means that the matrix M in Eq. (1.6) has to be $M = M_1 + \beta^2 M_2$, where

$$M_1 = \begin{pmatrix} b_1 & b_0 & 0 & \cdots & 0 & 0 & 0 \\ b_3 & b_2 & b_1 & \cdots & 0 & 0 & 0 \\ b_5 & b_4 & b_3 & \cdots & 0 & 0 & 0 \\ \vdots & \vdots & \vdots & \ddots & \vdots & \vdots & \vdots \\ 0 & 0 & 0 & \cdots & b_{n-2} & b_{n-3} & b_{n-4} \\ 0 & 0 & 0 & \cdots & 0 & 0 & b_{n-2} \\ 0 & 0 & 0 & \cdots & 0 & 0 & 0 \end{pmatrix},$$

$$M_2 = \begin{pmatrix} 0 & 0 & 0 & \cdots & 0 & 0 & 0 \\ b_1 & b_0 & 0 & \cdots & 0 & 0 & 0 \\ b_3 & b_2 & b_1 & \cdots & 0 & 0 & 0 \\ \vdots & \vdots & \vdots & \ddots & \vdots & \vdots & \vdots \\ 0 & 0 & 0 & \cdots & b_{n-4} & b_{n-5} & b_{n-6} \\ 0 & 0 & 0 & \cdots & b_{n-2} & b_{n-3} & b_{n-4} \\ 0 & 0 & 0 & \cdots & 0 & 0 & b_{n-2} \end{pmatrix}.$$

Let us note that the rth row of the matrix M_2 is equal to the $(r-1)$th row of matrix M_1 for $r > 1$. This means that all minors $\Delta_j(P)$, $j = 1, \ldots, n-2$, of the matrix M are equal to the corresponding minors $\Delta_j(Q)$ for the matrix M_1. The latter minors are the minors of the Hurwitz matrix related to the polynomial Q; so they are positive. We find also that $\Delta_{n-1}(P) = 0$ and $\Delta_n(P) = \beta^2 b_{n-2}\Delta_{n-1}(P) = 0$.

We see that the equation $\Delta_{n-1}(P) = 0$ describes locally the hypersurface in the coefficients space (a_j) separating the stable polynomials from the unstable ones. We only have to check whether the inequality $\Delta_{n-1}(P) > 0$ locally defines the set of stable polynomials.

For this purpose, we consider the following deformation of the situation (1.8):

$$P_{\alpha,\beta} = (\lambda^2 + \alpha^2\lambda + \beta^2)Q(\lambda),$$

where parameter α is small and β is real. Then there appears another summand $\alpha^2 M_3$ in the matrix M. In the last two rows of the matrix M, we have

$$0 \quad \ldots 0 \quad \beta^2 b_{n-2} \quad \alpha^2 b_{n-2} + \beta^2 b_{n-3} \quad a_{n-2}$$
$$0 \quad \ldots 0 \quad \quad 0 \quad \quad \quad 0 \quad \quad \quad \quad \beta^2 b_{n-2}.$$

When α and β are non-zero, the polynomial $P_{\alpha,\beta}$ is stable; so $\Delta_j(P_{\alpha,\beta}) \neq 0$ for $j = 1, \ldots, n-1$.

Let us calculate the behavior of $\Delta_{n-1}(P_{\alpha,\beta})$ (its sign) as $\beta \to 0$ and $\alpha \neq 0$ is fixed; then $\Delta_n(P_{\alpha,\beta}) \to 0$, because $a_n = \beta^2 b_{n-2} \to 0$, but this does not bother us. It is easy to see that, for $\beta = 0$ and small non-zero α, the matrix M takes a block form, with blocks: M_{11} of dimension $(n-2) \times (n-2)$, M_{12} of dimension $(n-2) \times 2$, $M_{21} = 0$ of dimension $2 \times (n-2)$ and M_{22} as above. Since $\det M_{11} = \Delta_{n-2}(P_{\alpha,0})$ is close to $\Delta_{n-2}(P_{0,0}) = \Delta_{n-2}(Q) > 0$ (by inductive assumption), so also $\det M_{11} > 0$. Thus,

$$\Delta_{n-1}(P_{\alpha,0}) = \det M_{11} \cdot \alpha^2 b_{n-2} > 0.$$

Finally, note that for $n = 3$, we have $P(\lambda) = (\lambda^2 + \beta^2)(b_0\lambda + b_1)$ and one directly checks that $\Delta_2 = 0$.

From the above proof also the second statement of the theorem follows. $\quad\square$

Example 1.16 (Watt's governor). In Fig. 1.5, we present a scheme of the Watt's governor used in the 19th century in steam engines. This regulator consists of:

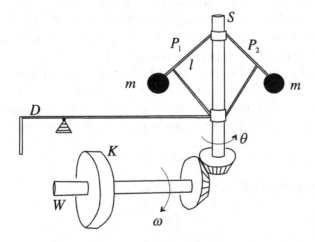

Fig. 1.5. Watt's governor.

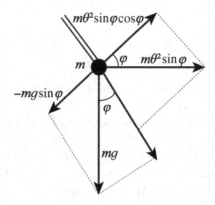

Fig. 1.6.   Gravity and centrifugal forces.

- a pin S, which can rotate around its axis;
- two balls of mass m each placed on movable joints around the pin S, so that the upper rim is fixed (merged with S) and the lower rim can move up and down (with the balls respectively moving away from the pin and approaching to the pin); furthermore, the rods P_1 and P_2 joining the balls from the upper rim have length l;
- a fly-wheel K placed on a roll W;
- a gearing between the pin S and the roll W with a speed ratio n;
- a lever D which regulates the steam supply to the machine and is attached to the lower rim.

There are three forces acting on each ball (see Fig. 1.6): the centrifugal force $F_{\text{cent}} = ml\theta^2 \sin \varphi$ (pointing perpendicularly from the pin to the outside), the gravity force $F_{\text{grav}} = mg$ (facing down) and the friction $F_{\text{fric}} = -b\dot{\varphi}$ (perpendicular to the bars $P_{1,2}$). Here θ is the speed of rotation of the pin S (and balls), φ is the angle between the rods $P_{1,2}$ and the pin S, g is the earth's acceleration and b is a coefficient. By summing-up the components of these forces perpendicular to the rods $P_{1,2}$, we get the following equation of motion:

$$ml\ddot{\varphi} = ml\theta^2 \sin \varphi \cos \varphi - mg \sin \varphi - b\dot{\varphi}. \qquad (1.9)$$

Usually it is assumed (e.g., in [3]) that

$$l = 1,$$

i.e., in certain units of length.

In Eq. (1.9), we have the additional variable θ, which also changes with time. To determine the dependency of θ (or its derivatives) on φ, let us first consider its relation

$$\theta = n\omega$$

with rotational speed ω of the cylinder W. On the other hand, the motion of the fly-wheel K is described by the equation

$$J\dot{\omega} = k\cos\varphi - F,$$

where J is the moment of inertia of the wheel, whereas on the right-hand side we have the moment of force acting on the wheel. In this case, the component $k\cos\varphi$ is proportional to the amount of steam (k is a constant) and F is the constant slowing force (due to the work done by the machine).

From the above, we arrive at the following closed and autonomous system of differential equations for $x = \varphi$, $y = \dot{\varphi}$ and $z = \omega$:

$$\dot{x} = y,$$
$$\dot{y} = n^2 z^2 \sin x \cos x - g \sin x - (b/m)\, y, \qquad (1.10)$$
$$\dot{z} = (k/J)\cos x - F/J.$$

It turns out that this system has exactly one (physically achievable) equilibrium position (x_0, y_0, z_0) given by the equations

$$\cos x_0 = F/k, \quad y_0 = 0, \quad n^2 z_0^2 = g/\cos x_0. \qquad (1.11)$$

Moreover, the linearization matrix of system (1.10) at this point of equilibrium is

$$A = \begin{pmatrix} 0 & 1 & 0 \\ -g\sin^2 x_0/\cos x_0 & -b/m & 2g\sin x_0/z_0 \\ -(k/J)\sin x_0 & 0 & 0 \end{pmatrix} \qquad (1.12)$$

and its characteristic polynomial is

$$\det(A - \lambda) = -P(\lambda) = -\left\{ \lambda^3 + \frac{b}{m}\lambda^2 + g\frac{\sin^2 x_0}{\cos x_0}\lambda + 2kg\frac{\sin^2 x_0}{Jz_0} \right\}. \qquad (1.13)$$

We can see that the coefficients of the polynomial $P(\lambda)$ are positive, that is condition (i) of the Routh–Hurwitz theorem is satisfied.

By Example 1.14 (for $n = 3$), the sufficient condition for stability of the polynomial $P(\lambda)$ is the inequality (1.7), which in this case means

$$\frac{bJ}{m} > 2k\frac{\cos x_0}{z_0} = \frac{2F}{z_0} \tag{1.14}$$

(Exercise 1.29).

Introduce the quantity $\nu := z_0/2F = \omega_0/2F$, which has mechanical interpretation of the *machine unevenness*. So, the last inequality is a simple one,

$$bJ\nu/m > 1.$$

The following conclusions can be drawn:

- increasing the mass m of the balls aggravates the stability;
- decreasing the friction coefficient b aggravates the stability[c];
- reducing the moment of inertia J of the fly-wheel aggravates the stability;
- a similar effect is caused by the reduction of the machine's unevenness factor ν.

Remark 1.17. One can ask what to do when neither the Lyapunov theorem nor the Chetaev theorem decide about the stability of an equilibrium point. In fact, this is a subject of the sequel chapters of this book. Anyway, we present some hints here.

In the planar case ($n = 2$), we have usually two situations: either (i) we have a pair of pure imaginary eigenvalues, or (ii) one eigenvalue vanishes (and the other is negative).

In case (i), if we somehow know that there exists a local regular first integral, then we have the Lyapunov stability (but not the asymptotic stability). One can also try to construct a Lyapunov function directly. Also one can check whether the divergence of the vector field has constant sign in a punctured neighborhood of the equilibrium point; by Theorem 6.24 in Appendix then the phase flow decreases or increases area.

---

[c]When the first author presented this example a few years ago at a lecture on the Qualitative Theory Z. Nowak informed the audience about cases when in some German factories (where everything was taken care of) a persistent decrease in the friction coefficient led to the failure of steam engines.

In case (ii), it is convenient to find the so-called center manifold and then the restriction of the vector field to it allows to decide the stability (see Section 3.3 in Chapter 3 for more details).

1.2. Hyperbolicity

The results of the previous section have taught us that the condition $\operatorname{Re} \lambda_j = 0$, for a certain eigenvalue of the linearization matrix A at the point of equilibrium of the autonomous vector field

$$\dot{z} = Az + \cdots , \quad z \in (\mathbb{R}^n, 0) \tag{1.15}$$

is a limiting condition for solving the asymptotic stability problems of this equilibrium point. So, there arises the following definition.

Definition 1.18. The equilibrium point $z = 0$ of the autonomous vector field (1.15) is called **hyperbolic**, if the real parts of all the eigenvalues of the linearization matrix A of the field at this point are non-zero.

Let us assume that the point $z = 0$ is hyperbolic and consider the corresponding linear system

$$\dot{z} = Az. \tag{1.16}$$

Then there is a natural decomposition of the space $\mathbb{R}^n$ into a direct sum of the so-called stable subspace $E^s \simeq \mathbb{R}^k$ and the unstable subspace $E^u \simeq \mathbb{R}^l$, corresponding to the values of $\operatorname{Re} \lambda_j < 0$ and $\operatorname{Re} \lambda_j > 0$, respectively:

$$\mathbb{R}^n = E^s \oplus E^u, \quad A = A^s \oplus A^u. \tag{1.17}$$

Let us note that the subspaces of E^s and E^u can be defined topologically in terms of the linear phase flow $g_{Az}^t = e^{At}$ (see Appendix) of the linear field (1.16). Namely,

$$E^s = \left\{ z : g_{Az}^t(z) \to 0, \ t \to +\infty \right\}, \quad E^s = \left\{ z : g_{Az}^t(z) \to 0, \ t \to -\infty \right\}$$

(see Fig. 1.7).

It turns out that the analogous situation occurs in the case of the nonlinear field (1.17).

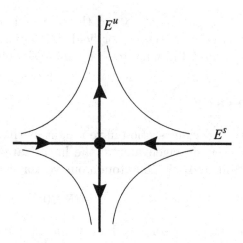

Fig. 1.7.   Hyperbolic saddle.

Theorem 1.19 (Hadamard–Perron). *For the hyperbolic equilibrium point $z = 0$ of the field $\dot{z} = v(z)$ of class C^r, $r > 2$, there exist local submanifolds, stable W^s and unstable W^u, of class C^r, such that*

$$W^s = \left\{z : g_v^t(z) \to 0,\ t \to +\infty\right\}, \quad W^s = \left\{z : g_v^t(z) \to 0,\ t \to -\infty\right\}, \tag{1.18}$$

and[d]

$$T_0 W^s = E^s, \quad T_0 W^u = E^u. \tag{1.19}$$

Before starting the proof of this theorem, let us note that analogous concepts and theorems can be introduced for **local diffeomorphisms**. First, if $z = 0$ is the equilibrium point of the vector field $\dot{z} = v(z) = Az + \cdots$, then $z = 0$ is a **fixed point** of the phase flow at time $t = 1$, $f(z) = g_v^1(z)$, i.e.,

$$f(0) = 0.$$

In addition, the linear part $\frac{\partial f}{\partial z}(z)$ of the transformation f at $z = 0$ has the form of the matrix

$$\partial f / \partial z(0) = B = e^A$$

---

[d]Here g_v^t denotes the local phase flow generated by the field $v(x)$ and $T_y M$ denotes the tangent space to the manifold M at the point y.

(Exercise 1.43). In fact, there is a discrete version of the concept of the phase flow.

Definition 1.20. A **diffeomorphism** $f : M \longmapsto M$ (of a manifold) defines the homomorphism $\mathbb{Z} \to \text{Diff}(M)$ from the additive group of integers to the group of diffeomorphisms of the manifold M, so that

$$n \longmapsto f^n,$$

where $f^n = f \circ \cdots \circ f$ (n times for $n \geq 0$) and $f^{-n} = f^{-1} \circ \cdots \circ f^{-1}$ ($|n|$ times for $n < 0$). In the literature, $\{f^n\}_{n \in \mathbb{Z}}$ is called the **cascade**.

A point $z_0 \in M$ is a **periodic point with period** $p \geq 1$ for f if $f^p(z_0) = z_0$; here (and in the following) we will understand the minimal period (i.e., $f^q(z_0) \neq z_0$ for $1 \leq q < p$). Of course, the periodic point with period $p = 1$ is a fixed point.

Definition 1.21. The periodic point z_0 of period p of a diffeomorphism f is called **hyperbolic** if the matrix

$$B = \partial(f^p)/\partial z(z_0)$$

has all its eigenvalues outside the unit circle,

$$|\lambda_j| \neq 1.$$

Lemma 1.22. *If $z = 0$ is a hyperbolic equilibrium point for the vector field $v(z)$, then $z = 0$ is also a hyperbolic fixed point of the diffeomorphisms $f = g_v^t$, $t \neq 0$, and vice versa (Exercise 1.43).*

We have the following version of the *Hadamard–Perron theorem for diffeomorphisms.*

Theorem 1.23. *If the fixed point $z = 0$ of a local diffeomorphism f : $(\mathbb{R}^n, 0) \longmapsto (\mathbb{R}^n, 0)$ of class C^r, $r \geq 1$, is hyperbolic then there exist local subspaces, stable W^s and unstable W^u, of class C^r, such that*

$$W^s = \{z : f^n(z) \to 0, \ n \to +\infty\}, \quad W^s = \{z : f^n(z) \to 0, \ n \to -\infty\}, \tag{1.20}$$

and

$$T_0 W^s = E^s, \quad T_0 W^u = E^u, \tag{1.21}$$

where E^s and E^u are the subspaces of $\mathbb{R}^n$ corresponding to the eigenvalues of the matrix $B = \frac{\partial f}{\partial z}(0)$ of modulus < 1 and > 1, respectively.

The path to the proof of the Hadamard–Perron Theorem 1.19 leads through the proof of Theorem 1.23. As we shall see, the method of finding the submanifolds W^s and W^u with properties (1.20) of class C^0 is quite natural: we arrive at a fixed point condition for a certain transformation in a corresponding infinite dimensional Banach space. Unfortunately, the "squeezing" of the contracting condition of this transformation is very exhaustive. Therefore, in the following proof, we will limit ourselves to derivation of corresponding equations and to sketching of the general estimation scheme. For a complete proof, we refer the reader to the monograph [4] of W. Szlenk.

Proof. [Proof of Theorem 1.23] To simplify the situation, let us assume decomposition (1.17), that is, $\mathbb{R}^n = E^s \oplus E^u = \{(x, y)\}$, and the transformation in the form $f = (f_1, f_2)$, such that

$$f_1(x, y) = Ax + \varphi(x, y), \quad f_2(x, y) = By + \psi(x, y), \qquad (1.22)$$

where

$$\|A\| < 1, \quad \|B^{-1}\| < 1 \qquad (1.23)$$

and the functions φ and ψ are of order in $o(|x| + |y|)$ (Exercise 1.44).

Of course, the vector functions φ and ψ are defined in a small neighborhood of zero. In the following proof, this is a technical obstacle. Therefore, we perform the following change:

$$\varphi \longmapsto \chi\varphi, \quad \psi \longmapsto \chi\psi,$$

where $\chi(x, y)$ is a smooth function (of class C^∞), non-negative and such that:

(i) $\chi(x, y) \equiv 1$ in a small neighborhood of zero, $|x| + |y| < \varepsilon$;
(ii) $\chi(x, y) \equiv 0$ outside the small neighborhood of zero, $|x| + |y| > 2\varepsilon$ (Exercise 1.45).

Thus, the vector-valued functions $\chi\varphi$ and $\chi\psi$, after the zero extension for $|x| + |y| > 2\varepsilon$, will be defined on all $\mathbb{R}^n$; in the sequel we still denote them by φ and ψ. Recall that these new functions satisfy $\varphi(0, 0) = 0$, $d\varphi(0, 0) = 0$, $\psi(0, 0) = 0$, $d\psi(0, 0) = 0$, and $|\varphi|$ and $|\psi|$ are small with derivatives. By property (i) the dynamics of the transformation f, with the new φ and ψ, in a neighborhood of the origin is the same as of the old transformation (1.22).

We are looking for the submanifold W^s in the form of the graph of certain mapping (or vector function) $F : E^s \longmapsto E^u$,

$$W^s = \{(x, F(x)) : x \in E^s\}.$$

(Proof of the existence of the submanifold W^u is quite similar, so we limit ourselves to the case of W^s.)

From property (1.20) it follows that the submanifold W^s should be invariant for the diffeomorphism f, $f(W^s) = W^s$. This means that $f(x, F(x)) = (x_1, F(x_1))$ for some $x_1 \in E^s$ depending on $x \in E^s$. From Eq. (1.22), we find that $x_1 = Ax + \varphi(x, F(x))$. So, we get the condition

$$BF(x) + \psi(x, F(x)) = F \circ (Ax + \varphi(x, F(x)),$$

which we will write in the following form:

$$F(x) = B^{-1} \left\{ F \circ (Ax + \varphi(x, F(x)) - \psi(x, F(x)) \right\} =: \mathcal{T}(F)(x).$$
$$(1.24)$$

We treat the latter equation as a fixed point equation $F = \mathcal{T}(F)$ for the nonlinear operator $\mathcal{T}$ defined above.

Assuming that the functions φ and ψ are of class C^1, it is natural to introduce the Banach space $\mathcal{X} = C^0(E^s, E^u)$ of continuous mappings with the supremum norm. It is easy to show that the transformation $\mathcal{T}$ sends $\mathcal{X}$ into itself. In order to apply the Banach theorem for contacting mappings, it is still necessary to prove the contraction condition, i.e., to estimate the norm of the difference $\mathcal{T}(F_1) - \mathcal{T}(F_2)$. Here a problem arises, because from Eq. (1.24) we get the following inequality:

$$\|\mathcal{T}(F_1) - \mathcal{T}(F_2)\| \leq \|B^{-1}\| \left\{ 1 + \|F_1'\| \cdot \|\varphi_y'\| + \|\psi_y'\| \right\} \cdot \|F_1 - F_2\|$$

(Exercise 1.46). Since $\|B^{-1}\| < 1$ (see Eq. (1.23)) and $\|\psi_y'\| = \|\partial\psi/\partial y\|$ and $\|\partial\varphi/\partial y\|$ are small (see above), it remains only to estimate the norm of the derivative F_1' of the mapping F_1. But, if we choose F_1 and F_2 arbitrarily from space $\mathcal{X}$, then F_1 will be only continuous, and its derivative can be unbounded.

There is an outcome of this deadlock. Let us recall that in the proof of Banach's theorem one chooses $F_0 \in \mathcal{X}$ and then $F_n = \mathcal{T}^n(F_0)$ should converge to a fixed point. The point is to choose the vector function F_0 in a special way which allows us to show that the functions F_n are also smooth with appropriately bounded norms. It is easy to guess that

$$F_0(x) \equiv 0$$

is a good choice. It is also easy to see from formula (1.24) that $F_n(x)$ are smooth, e.g., $F_1(x) = -B^{-1}\psi(x,0)$.

You just have to show that the functions $F_n(x)$ are uniformly continuous. This is reduced to the estimation of the norm of the derivative $(\mathcal{T}(F))'(x)$ under the assumption of a bound for the norm of $F'(x)$. We have

$$(\mathcal{T}(F))'(x) = B^{-1} \cdot \{F' \cdot [A + \varphi'_x + \varphi'_y \cdot F'] - \psi'_x - \psi'_y \cdot F'\}, \quad (1.25)$$

where we omitted the arguments of the functions appearing on the right side of this equality. Thus, the supremum norm is estimated as follows:

$$\|(\mathcal{T}(F))'\| \le a + b\|F'\| + c\|F'\|^2,$$

where a is small, $b < 1$ and $c > 0$. Hence, if $\|F'\|$ is small enough, $\|F'\| < d$ (for appropriate d), then $\|(\mathcal{T}(F))'\| < d$ (Exercise 1.47). This gives a uniform estimate for the norms $\|F'_n\|$ of the functions F_n.

So, F_n's converge to the fixed point F_*, about which we can only say that it is represented by a continuous mapping of E^s to E^u; that is the submanifold

$$W^s = \{(x, F_*(x))\}$$

is of class C^0.

Let us briefly explain how to demonstrate the smoothness of the function F_*. To do this, Eqs. (1.24)–(1.25) must be applied simultaneously to the series $\{F_n\}$ and $\{F'_n\}$. In particular, we show the uniform continuity of the family $\{F'_n\}$, which requires a uniform estimation of the expression $\sup |(\mathcal{T}(F_n))'(x_1) - (\mathcal{T}(F_n))'(x_2)|$. It turns out that this can be done using the estimates for $\sup \max\{|F'_n(x_1) - F'_n(x_2)|, |\varphi'(x_1, y_1) - \varphi'(x_2, y_2)|, |\psi'(x_1, y_1) - \psi'(x_2, y_2)|\}$.

Then we use Ascoli theorem, which says that from a sequence of uniformly continuous functions on a compact set one can choose a convergent subsequence. Here the compact set is $\{|x| < M\} \subset E^s$ for some M and the limit of $\{F_{n_k}\}$ must be F_* (because it is the limit in the space of continuous functions).

In this (shortened) proof, we have limited ourselves to the case when f is of class C^1 (and then $W^{s,u}$ are also classes C^1). But the case of the class C^r for $r > 1$ can be done by the same method, only the proof requires a larger number of formulas and estimates. We skip it.

Finally, we note that, since $F_0'(0) = 0$ and $\varphi'(0,0) = 0$ and $\psi'(0,0) = 0$, then we have $F_*'(0) = 0$, which means that the submanifold W^s is tangent to $(0,0)$ to the space E^s. □

Proof of Theorem 1.19. Let us assume $f = g_v^1$, that is the time one phase flow, and let V^s be a local stable variety for f (see Theorem 1.23). Since the submanifold W^s is defined topologically as a set of these points z such that $g_v^t(z) \to 0$ when $t \to \infty$, then $W^s \subset V^s$. On the other hand, if $z \in V^s$, by writing $t = n + \tau$ for $n \in \mathbb{N}$ and $0 \le \tau < 1$, we have $g^t(z) = g^\tau(g^n(z)) \to 0$ (as the family $\{g^\tau\}_{\tau \in [0,1)}$ is uniformly continuous). □

The second basic result about hyperbolic points comes from D. Grobman and P. Hartman [5]. We formulate it simultaneously for cascades and flows.

Theorem 1.24 (Grobman–Hartman). *Let $f \colon (\mathbb{R}^n, 0) \longmapsto (\mathbb{R}^n, 0)$ be a diffeomorphism of class C^r, $r \ge 1$, with a hyperbolic fixed point at $z = 0$. Then there exists a local homeomorphism $h : (\mathbb{R}^n, 0) \longmapsto (\mathbb{R}^n, 0)$ such that*

$$h \circ f'(0)(z) = f \circ h(z). \qquad (1.26)$$

Analogously, for a local flow g_v^t generated by a vector field $v(z)$ with a hyperbolic equilibrium point at $z = 0$, there exists a local homeomorphism h (as above) such that

$$h \circ e^{tv'(0)}(z) = g_v^t \circ h(z). \qquad (1.27)$$

Proof. Let us start with the cascade case. Similarly to the proof of Theorem 1.23, we reduce the situation to the case where $z = (x, y)$ and

$$f(x,y) = (Ax + \varphi, By + \psi) = Lz + \tilde{f},$$

where $L = A \oplus B = f'(0)$, estimates (1.23) hold and $\tilde{f} = (\varphi(x,y), \psi(x,y))$ is defined on the whole space $\mathbb{R}^n = E^s \oplus E^u$ and is small with derivatives. The homeomorphism h will be chosen in the form

$$h = id + g = (x + g_1, y + g_2), \quad g \text{ small.} \tag{1.28}$$

Equation (1.26) on h, which means the commutativity of the diagram

$$
\begin{array}{ccc}
\mathbb{R}^n & \overset{f}{\longmapsto} & \mathbb{R}^n \\
\uparrow h & & \uparrow h \\
\mathbb{R}^n & \overset{L}{\longmapsto} & \mathbb{R}^n
\end{array}
$$

leads to the equation $(id + g) \circ L = L(id + g) + \tilde{f} \circ (id + g)$. In components, we get the following system of equations:

$$g_1(Ax, By) = Ag_1(x, y) + \varphi(x + g_1, y + g_2),$$
$$g_2(Ax, By) = Bg_2(x, y) + \psi(x + g_1, y + g_2).$$

Let us rewrite it in a more convenient way:

$$
\begin{aligned}
g_1(x, y) &= Ag_1(A^{-1}x, B^{-1}y) + \varphi \circ (id + g) \circ (A^{-1}x, B^{-1}y), \\
g_2(x, y) &= B^{-1}g_2(Ax, By) - B^{-1} \cdot \psi \circ (id + g).
\end{aligned}
\tag{1.29}
$$

It is easy to recognize here the fixed point equation, $g = \mathcal{T}(g)$, for a nonlinear operator $\mathcal{T}$ acting on $g = (g_1, g_2)$ by the right-hand sides of Eq. (1.29).

We take the Banach space

$$\mathcal{X} = C^0(\mathbb{R}^n, E^s) \oplus C^0(\mathbb{R}^n, E^u)$$

with the norm $\|g\| = \sup |g_1| + \sup |g_2|$. It is easy to show that the operator $\mathcal{T}$ sends a ball in $\mathcal{X}$ with a suitable radius to itself and is a contraction. The basic argument is that the matrices A and B^{-1} have a norm < 1.

Let us stop our proof for a moment and consider the situation when Eq. (1.26) is replaced with the equation

$$k \circ f = L \circ k, \tag{1.30}$$

where $k : \mathbb{R}^n \longmapsto \mathbb{R}^n$. After the substitution $k = id + l = (x + l_1, y + l_2)$ and some manipulations, we get the following system (analogous to

Eq. (1.29)):

$$l_1(x, y) = A l_1 \circ f^{-1}(x, y) - \varphi \circ f^{-1}(x, y),$$
$$l_2(x, y) = B^{-1} l_2(x, y) + B^{-1} \psi(x, y).$$

Here we are dealing with a fixed point equation for the corresponding transformation $\mathcal{S} : \mathcal{X} \longmapsto \mathcal{X}$, which is contractible. So, also system (1.30) has a solution.

Let us note the following property of the solutions to Eqs. (1.26) and (1.28), which are the consequence of the fact that, in the Banach fixed point theorem, the fixed point depends continuously on parameters (if the transformation itself depends on parameters continuously):

The solutions $h(x, y)$ and $k(x, y)$ of Eqs. (1.26) and (1.30) are unique and depend in a continuous way on the data in these equations (i.e., $L = A \oplus B$ and $\tilde{f} = (\varphi, \psi)$). In addition, in Eq. (1.26) we can replace the linear transformation $L = f'(0)$ by any transformation e such that $e'(0) = L$.

The above-mentioned uniqueness will allow us to prove that the transformations h and k are homeomorphisms; more precisely, that $h \circ k = k \circ h = id$. In fact, the transformation $m = k \circ h$ satisfies the condition $m \circ L = L \circ m$, i.e., Eq. (1.26) for $f = L$. Analogously, the transformations $n = h \circ k$ and id satisfy the equation $f \circ n = n \circ f$.

Let us now pass to the proof of the second part of the theorem, that is, that the homeomorphism of h satisfies all equations (1.27) for the family of transformations $f_t = g_v^t$, $v = Lz + \cdots$.

For each $t \neq 0$, the transformation f_t has the hyperbolic fixed point $z = 0$. Thus, from the proved first part of the theorem, we have the existence of a family of homeomorphisms h_t, $t \neq 0$, such that

$$h_t \circ e^{Lt} = f_t \circ h_t.$$

We just have to show that h_t does not depend on t, which we treat here as a parameter. But we know that h_t depends on t in a continuous way.

Consider the following identity:

$$h_{t/2}^{-1} \circ f_t \circ h_{t/2} = (h_{t/2}^{-1} \circ f_{t/2} \circ h_{t/2}) \circ (h_{t/2}^{-1} \circ f_{t/2} \circ h_{t/2}) = e^{Lt/2} \circ e^{Lt/2} = e^{Lt}$$

(here we used the group property of the phase flow). It means that $h_{t/2} = h_t$ (uniqueness). Similarly, it is proved that $h_{t/k} = h_t$ for natural k and hence that

$$h_{(k/l)t} = h_t, \quad k, l \in \mathbb{N}$$

(Exercise 1.48). Hence, for the set of rational parameters t the transformations h_t are the same. The continuous dependence of h_t on the parameter (see above) implies $h_t \equiv h = $ const as a function of $t > 0$. Now, the observation that, if h satisfies Eq. (1.27) for a given time $t > 0$ then also for for the time $-t$ (Exercise 1.49), finishes the proof.

Finally, one more note. Since $\{g_v^t\}$ is only a local phase flow (for the vector field $v(z)$ defined in the neighborhood of $z = 0$), it is necessary to take care of the domains of definitions of the flow maps and, thus, the domains of transformations h_t. But there is no problem here, because the domain of transformation $g_v^{t/k}$ increases with the increase of $k \in \mathbb{N}$. Just, in the above proof, we limit ourselves to the times such that $|t| < 1$.   □

Equation (1.26) means that the dynamics (i.e., the cascade) generated by the diffeomorphism f is the same from the qualitative point of view as the dynamics generated by the linear diffeomorphism $L(z) = f'(0)z$. Indeed, if $\text{Orb}_f(z_0) = \{\ldots, f^{-1}(z_0),$ $z_0, f(z_0), f^2(z_0), \ldots\}$ is the **orbit of the point** z_0 under the diffeomorphism f and $y_0 = h(x_0)$, then $h(\text{Orb}_f(z_0))$ is the orbit $\text{Orb}_L(y_0) = \{\ldots, L^{-1}(y_0), y_0, L(y_0), \ldots\}$ of the point y_0 under to the linear diffeomorphism L.

The following definition seems natural.

Definition 1.25. If for the diffeomorphisms $f : M \longmapsto M$ and $g : N \longmapsto N$ there exists a homeomorphism $h : M \longmapsto N$ such that

$$g = h \circ f \circ h^{-1},$$

then we say that the diffeomorphism f and g are **topologically conjugated** (via h). If h is of class C^r, then we say that f and g are C^r — **conjugated**. Similarly, the vector fields $v(x)$ and $w(x)$ are topologically (or of classes C^r) conjugated if their phase flows are conjugated by a homeomorphism (or a diffeomorphism of class C^r).

If the diffeomorphism f has the property that any diffeomorphism g which is close (in a sense that we do not want to specify here) to f is topologically conjugated with f, then we say that f is **structurally stable**. Similarly, the vector field $v(x)$ is **structurally stable** if the near vector fields are topologically conjugated to it.

Hartman–Grobman theorem states that a diffeomorphism (respectively, a vector field) in a neighborhood of a hyperbolic fixed point (respectively, a hyperbolic equilibrium point) is topologically conjugated to the linear part of the diffeomorphism (respectively, the vector field). We can prove more.

Proposition 1.26. *A diffeomorphism (respectively, a vector field) in a neighborhood of a hyperbolic fixed point (respectively, a hyperbolic equilibrium point) is structurally stable.*

Proof. We use the following direct construction of a homeomorphism h, which conjugates two local diffeomorphisms f and g in the case of asymptotically stability, i.e., when $f'(0)$ and $g'(0)$ have all their eigenvalues of modulus <1. We can assume that $E^s = \mathbb{R}^n$ and $z = x$ in the proof of Grobman–Hartman theorem. Then there exists a "Lyapunov function", $L(x)$, which satisfies condition (1) of Definition 1.5 and the following (analogous to (2)) condition:

$$L(f(x)) < L(x) \quad \text{for} \quad x \neq 0.$$

Its construction is exactly the same as in the proof of the Lyapunov theorem, we can assume that $L(x) = |x|^2$ in a corresponding (linear) coordinate system. Let $M(x) = |x|^2$ be the corresponding Lyapunov function for the diffeomorphism g (also in a corresponding coordinate system). We have two copies of $\mathbb{R}^n$, on which the diffeomorphisms f and g act, respectively.

Let us take a small $\varepsilon > 0$ and consider hypersurfaces (diffeomorphic with spheres) $\{L(x) = \varepsilon\}$ and $\{M(x) = \varepsilon\}$. Let us define the homeomorphism between these hypersurfaces as $h|_{L=\varepsilon} : \{L = \varepsilon\} \longmapsto \{M = \varepsilon\}$ as $id|_{\{L=\varepsilon\}}$ (see Fig. 1.8). The condition

$$h \circ f^{-1} = g^{-1} \circ h \tag{1.31}$$

allows us to "define" h between the hypersurfaces $f(\{L = \varepsilon\})$ and $g(\{M = \varepsilon\})$, as shown in Fig. 1.8. We prolong h to a continuous and

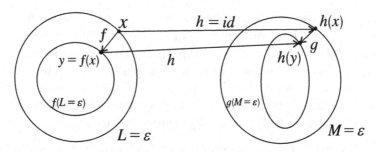

Fig. 1.8.   Conjugacy.

one-to-one map on the domain between the hypersurfaces $\{L = \varepsilon\}$ and $f(\{L = \varepsilon\})$. Using successively Eq. (1.31), we extend h to the whole domain $\{0 < L \leq \varepsilon\}$. By putting $h(0) = 0$, we get the desired homeomorphism.

Analogous construction works in the case of expanding diffeomorphisms, i.e., when the $f'(0)$ and $g'(0)$ have their eigenvalues with modulus >1.

Let us now consider two linear diffeomorphisms f_0 and g_0 defined by the hyperbolic matrices $A = A^s \oplus A^u$ and $B = B^s \oplus B^u$ in the corresponding (and the same) splitting $\mathbb{R}^n = E^s \oplus E^u$. From the above analysis, we get the homeomorphisms h^s and h^u, which conjugate $A^s x$ with $B^s x$ and $A^s y$ with $B^u y$, respectively. Now, the homeomorphism

$$h = h^s \oplus h^u$$

conjugates f_0 with g_0.

Let us now consider the diffeomorphism f in a neighborhood of the hyperbolic fixed point $z = 0$ and its small perturbation g with the same fixed point. Since the matrix $B = g'(0)$ is close to the matrix $A = f'(0)$, it is also hyperbolic with the same dimensions of the stable and unstable subspaces; that is, we can apply the above construction of the conjugating homeomorphism between the linear parts of these diffeomorphisms. We see that f is conjugated with $f_0 = f'(0)z$, f_0 is conjugated with $g_0 = g'(0)z$ and g_0 is conjugated with g; by comparing these three homeomorphisms one gets that f is conjugated with g.

The case of Proposition 1.26 for vector fields is left to the readers as an exercise (Exercise 1.50).   □

Remark 1.27. One can ask whether Grobman–Hartman's thesis can be strengthened, i.e., whether the homeomorphism h can be of class C^1. It turns out that no. For example, the transformation

$$(x, y, z) \longmapsto (\lambda x, \mu y, \lambda \mu (z + xy)), \quad 0 < \mu < 1 < \lambda, \quad \lambda \mu > 1, \quad (1.32)$$

cannot be linearized using a C^1-diffeomorphism.[e]

Indeed, suppose that there exists a change $x = \varphi(u, v, w)$, $y = \psi(u, v, w)$, $z = \chi(u, v, w)$ of class C^1 and linearizing map (1.32). Therefore, putting $w = 0$ and denoting $\varphi_0(u, v) = \varphi(u, v, 0)$, etc., we obtain the relations $\lambda \varphi_0(u, v) = \varphi_0(\lambda u, \mu v)$, $\mu \psi_0(u, v) = \psi_0(\lambda u, \mu v)$ and $\lambda \mu [\chi_0(u, v) + \varphi_0(u, v) \psi_0(u, v)] = \chi_0(\lambda u, \mu v)$. Iterating them we find the series of relations

$$\lambda^n \varphi_0 = \varphi_0(\lambda^n u, \mu^n v),$$

$$\mu^n \psi_0 = \psi_0(\lambda^n u, \mu^n v),$$

$$(\lambda \mu)^n (\chi_0 + n \varphi_0 \psi_0) = \chi_0(\lambda^n u, \mu^n v).$$

Putting $u = 0$ in the first of them and taking the limit as $n \to \infty$, we find that $\varphi_0(0, v) = 0$ (since $\lambda > 1$ and $\mu < 1$). Analogously, putting $v = 0$ and replacing u with u/λ^n in the second, we show that $\psi_0(u, 0) = 0$. Since $\lambda \mu > 1$, the third series for $u = 0$ implies $\chi_0(0, v) = 0$.

Let us rewrite the third series as follows:

$$\lambda^n \chi_0 (u/\lambda^n, v) + n \varphi_0(u, \mu^n v) \psi_0(u/\lambda^n, v) = (1/\mu)^n \chi_0(u, \mu^n v),$$

i.e., we have replaced u with u/λ^n. The first term on the left-hand side tends to $u \cdot (\partial \chi_0/\partial u)(0, v)$. The right-hand side equals $[\chi_0(u, \mu^n v) - \chi_0(u, 0)] \mu^n + (1/\mu)^n \chi_0(u, 0) \approx v \cdot (\partial \chi_0/\partial v)(u, 0) + (1/\mu)^n \chi_0(u, 0)$. Therefore, we obtain a relation of the approximate form

$$n \varphi_0(u, 0) \psi_0(0, v) \approx (1/\mu)^n \chi_0(u, 0),$$

where the "zeroes" mean u/λ^n and $\mu^n v$ respectively and the approximation $\approx$ is modulo $O(1)$. The only possible solution to the latter relation for $uv \neq 0$ is that $\chi_0(u, 0) = 0$ and $\varphi_0(u, 0) \psi_0(0, v) = 0$.

---

[e]This example comes from the work P. Hartman, A lemma in the theory of structural stability of differential equations, *Proc. Amer. Math. Soc.* 11 (1960), 610–620 (see also [5, Exercise 8.1]). Important property of this example is the resonance between the eigenvalues: $\lambda_3 = \lambda_1 \lambda_2$. This is the main obstacle to the smooth linearizability. Compare also Remark 3.23 in Chapter 3.

But, if $\varphi_0(u, 0)$ then we find that $\varphi = \psi = \chi = 0$ for $(u, v, w) = (u, 0, 0) \neq (0, 0, 0)$; this contradicts to the fact that the change is one-to-one. If $\psi_0(0, v) = 0$, then $\varphi = \psi = \chi = 0$ for $(u, v, w) = (0, v, 0) \neq (0, 0, 0)$.

This question about the smoothness of the conjugating map is related to resonances between eigenvalues (see Poincaré–Dulac Theorem 3.19 in Section 3.3).

1.3. Exercises

Exercise 1.28. Complete the proof of Lemma 1.7, i.e., in case of non-real eigenvalues.

Exercise 1.29. Prove formulas (1.10)–(1.14).

Exercise 1.30. Determine the stability (in the Lyapunov and the asymptotic sense) of the singular point $x = d/c$, $y = a/b$ of the **Lotka–Volterra system**

$$\dot{x} = x\,(a - by)\,, \quad \dot{y} = y\,(cx - d)\,, \quad a, b, c, d > 0, \qquad (1.33)$$

which describes the dynamics of two competing populations (predators and preys).
Hint: Remark 1.17 and Theorem 2.13.

Exercise 1.31. Using Definition 1.1 check that the equilibrium point $x = 0$ for the equation $\dot{x} = 4x - t^2 x$ is stable in Lyapunov's sense, i.e., with arbitrary t_0.

Exercise 1.32. Show that the zero solution to the equation $\dot{x} = a(t)x$ is Lyapunov stable if and only if $\limsup_{t \to +\infty} \int_0^t a\,(s)\,\mathrm{d}s < \infty$.

Exercise 1.33. Determine the stability of the equilibrium point $x = y = 0$ for the system $\dot{x} = \mathrm{e}^{x+2y} - \cos 3x$, $\dot{y} = \sqrt{4 + 8x} - 2\mathrm{e}^y$.

Exercise 1.34. Determine the stability of the zero solution of $\dot{x} = \mathrm{e}^x - \mathrm{e}^{3z}$, $\dot{y} = 4z - 3\sin(x + y)$, $\dot{z} = \ln(1 + z - 3x)$.

Exercise 1.35. Show the asymptotic stability of the point $(0, 0)$ for the vector field $\dot{x} = 2y - x - y^3$, $\dot{y} = x - 2y$.
Hint: Find a Lyapunov function as a homogeneous quadratic form.

Exercise 1.36. Let the functions $f_j(z)$, $j = 1, 2, 3, 4$, satisfy sign $f_j(z) = \text{sign } z$. Show that the origin is asymptotically stable for the system $\dot{x} = -f_1(x) + f_2(y)$, $\dot{y} = -f_3(x) - f_4(y)$.

Exercise 1.37. For which value of the parameter a, the zero solution of $\dot{x} = ax + y + x^2$, $\dot{y} = x + ay + y^2$ is asymptotically stable?
Hint: When $a = -1$ the straight line $y = x$ is invariant.

Exercise 1.38. For which parameters a and b, the zero solution of $\dot{x} = y + \sin x$, $\dot{y} = ax + by$ is asymptotically stable?
Hint: For $a = b \leq -1$, by introducing $z = \dot{x}$, the system becomes $\dot{x} = H'_z$, $\dot{z} = -H'_x + (a + \cos x)z$, $H = \frac{1}{2}z^2 - a\left(x^2/2 + \cos x\right)$; find a Lyapunov function.

Exercise 1.39. For which a the zero solution of $\dot{x} = ax - 2y + x^2$, $\dot{y} = x + y + xy$ is stable?
Hint: Try the variables $x, u = y/x$.

Exercise 1.40. Show that the origin is Lyapunov stable for the system $\dot{x} = y$, $\dot{y} = -x + x^3y$.
Hint: Remark 2.15.

Exercise 1.41. For which parameters of a and b, the solution $x(t) \equiv 0$ of the equation $\dddot{x} + 3\ddot{x} + a\dot{x} + bx = 0$ is asymptotically stable?

Exercise 1.42. Is the zero solution to the equation $x^{(5)} + 2x^{(4)} + 4x^{(3)} + 6x^{(2)} + 5x^{(1)} + 2x = 0$?

Exercise 1.43. Show that the diffeomorphism g^t of the (local) phase flow generated by the vector field $\dot{x} = Ax + O(|x|^2)$ has the linear part at the fixed point $x = 0$ of the form $B = e^{At}$. Deduce from this Lemma 1.22.
Hint: Theorem 6.7 from Appendix.

Exercise 1.44. Prove estimates (1.23) (for a corresponding coordinate system and Euclidean norm in $\mathbb{R}^n$).

Exercise 1.45. Give some explicit formula for the function χ from proof of Theorem 1.23.

Exercise 1.46. Prove the inequalities for $\|\mathcal{T}(F_1) - \mathcal{T}(F_2)\|$ from proof of Theorem 1.23.

Exercise 1.47. Find the constant d, depending on a, b, c, in the inequalities $\|F'\| < d$ and $\|(\mathcal{T}(F))'\| \le d$ from proof of Theorem 1.23.

Exercise 1.48. Prove that $h_{(k/l)t} = h_t$ for $k, l \in \mathbb{N}$ and $t \ne 0$ in the proof of Theorem 1.24.

Exercise 1.49. Prove that, if h satisfies property (1.27) for a given $t > 0$, then it also satisfies this property for $t < 0$.

Exercise 1.50. Complete the proof of Proposition 1.26.

Exercise 1.51. Check that the maps $(u, v, w) = (x, y, z - xy \ln |x| / \ln \lambda)$ and $(u, v, w) = (x, y, z - xy \ln |y| / \ln \mu)$ linearize map (1.32). Are they differentiable?

Chapter 2

Phase Portraits of Vector Fields

Definition 2.1. The **phase portrait** of an autonomous vector field $v(x)$ on a manifold M is the partition of M into the phase curves of this field.

Phase curves are of three types:

(i) equilibrium points, i.e., degenerate curves corresponding to constant solutions;

(ii) embedded intervals (bounded or not), i.e., the images $\varphi(I)$, $I \subset \mathbb{R}$, of the solutions of $\varphi : I \longmapsto M$ which are embeddings;

(iii) **closed phase curves** (embedded circles) corresponding to **periodic solutions** φ:

$$\varphi(t + T) = \varphi(t), \quad t \in \mathbb{R},$$

where $T > 0$ is the **period** of the solution (we assume that this is the minimum of T's satisfying this property).

Throughout this chapter, we consider only autonomous vector fields; therefore, we will skip the adjective "autonomous".

Example 2.2 (Mathematical pendulum). By splitting the gravity force into the component along the pendulum and a component orthogonal to it, i.e., $-mg \sin x$, the corresponding Newtonian

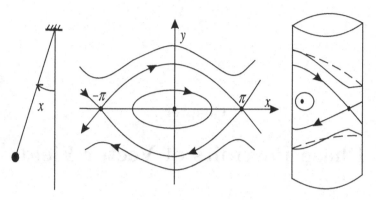

Fig. 2.1.   Pendulum.

equation is $ml\ddot{x} = -mg\sin x$ (see Fig. 2.1). After rescaling the time the corresponding first order system is

$$\dot{x} = y, \quad \dot{y} = -\sin x.$$

It is a system on the phase space $M = \mathbb{S}^1 \times \mathbb{R}$ (cylinder). It is easy to see that the function

$$H = y^2/2 - \cos x$$

is the first integral of this system, i.e., $\dot{H} \equiv 0$. Note the following properties of H:

- the equilibrium point $(0,0)$ is an absolute minimum point and $H(0,0) = -1$;
- the equilibrium point $(\pi,0)$ is a saddle point and $H(\pi,0) = 1$;
- $H(x,y) \to \infty$ as $|y| \to \infty$.

It is also easy to see that, in addition to the above equilibrium points, we have two phase curves of type (ii); these are the separatrices of the saddle point $(\pi,0)$ lying in the level $\{H = 1\}$. The remaining phase curves are closed and can be divided into two groups: (a) around the equilibrium point $(0,0)$ (corresponding to oscillations of limited amplitude) and (b) circulating the cylinder (they correspond to pivoting movements).

We can calculate the *periods* of the above periodic solutions lying on the level $H = h$ of the first integral. We have $\mathrm{d}t = \mathrm{d}x/y$, where

$y = \pm\sqrt{2(h + \cos x)}$. So, in case (a), we have

$$T = 2 \int_{x_1}^{x_2} \frac{dx}{\sqrt{2(h + \cos x)}},$$

where $x_{1,2}$ are two zeros of function $h + \cos x$. Here the integral from x_1 to x_2 equals the time the trajectory spends in the domain $y > 0$, but, due to symmetry, it is exactly half the period. In case (b), we have

$$T = \int_0^{2\pi} \frac{dx}{\sqrt{2(h + \cos x)}}.$$

Unfortunately, the above integrals cannot be expressed in terms of elementary functions. In fact, after substituting $u = \cos x$ (with $dx = du/\sin x = -du/\sqrt{1 - u^2}$) we get

$$T = 4 \int_{-h}^{1} \frac{du}{\sqrt{(1 - u^2)(h + u)}}.$$

The right side of the last equation is an *elliptic integral* defining certain elliptic function[a] (Exercise 2.50).

We also note that the closed phase curves in this example are not isolated, they form whole families.

2.1. Periodic solutions and focus quantities

Closed phase curves are also called periodic trajectories or periodic orbits. In Example 2.2, they arise in whole families, but there exist situations with isolated periodic trajectories.

Definition 2.3. An isolated closed phase curve of an autonomous vector field is called the **limit cycle**.

If the phase space is two-dimensional, $\dim M = 2$, then an equilibrium point of such a field, which is surrounded by non-isolated closed phase curves, is called the **center**.

---

[a]Elliptic integrals and elliptic functions appear very often in differential equations of classical mechanics (see [6]).

Example 2.4. Let us consider the system

$$\dot{x} = x(1 - x^2 - y^2) + y, \quad \dot{y} = -x + y(1 - x^2 - y^2).$$

It is convenient to rewrite it in the polar coordinate system r, φ:

$$\dot{r} = r(1 - r^2), \quad \dot{\varphi} = -1$$

(Exercise 2.52). We see that the solutions starting with $r = r_0 \in (0, 1)$ grow with time to $r = 1$ and solutions starting with $r_0 > 1$ are decreasing to $r = 1$. The solution starting from $r_0 = 1$ is constant and corresponds to the isolated periodic solution on the XY plane (see Fig. 2.2).

Definition 2.5. Let γ be a closed phase curve of a vector field in M. Take a germ S (a **section**) of a hypersurface transversal (i.e., at a non-zero angle) to γ at a point $p_0 \in \gamma$. From points $x_0 \in S$ start solutions $\varphi(t; x_0)$ which, after some times $T(x_0)$, return to S, $\varphi(T(x_0); x_0) \in S$. The resulting mapping $f : S \longmapsto S$ (a diffeomorphism on a corresponding domain),

$$x_0 \longmapsto f(x_0) = \varphi(T(x_0); x_0)$$

is called the **Poincaré return map** (see Fig. 2.3).

There is a certain arbitrariness in this definition, associated with the choice of the section S. It turns out that this is not a big problem because, if $f' : S' \longmapsto S'$ is the return transformation associated with another slice S', then we have the following lemma.

Lemma 2.6. *The diffeomorphisms f and f' are conjugated by a certain diffeomorphism of the same class of smoothness as f and f'.*

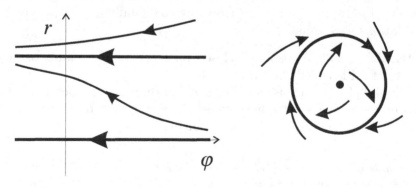

Fig. 2.2.   Limit cycle.

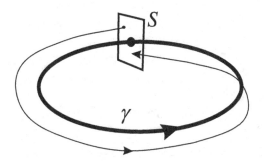

Fig. 2.3.   Return map.

Proof.   Let $f_1 : S \longmapsto S'$ and $f_2 : S' \longmapsto S$ be the natural maps along the solution. We have $f = f_2 \circ f_1$ and $f' = f_1 \circ f_2$.   □

The section (S, p_0) can be identified with $(\mathbb{R}^{n-1}, 0)$, where $n = \dim M$, and the return map defines a germ of diffeomorphism $f : (\mathbb{R}^{n-1}, 0) \longmapsto (\mathbb{R}^{n-1}, 0)$ (because $f(p_0) = p_0$) of the form

$$f(z) = Az + \cdots$$

(Exercise 2.53).

Definition 2.7. The closed phase curve γ is **hyperbolic** if the fixed point $z = 0$ for the above diffeomorphism is hyperbolic, i.e., $|\lambda_j| \neq 1$ for the eigenvalues of matrix A.

The following two propositions are direct analogues of the Lyapunov theorem and the Hadamard–Perron theorem.

Proposition 2.8. *If $|\lambda_j| < 1$ for all eigenvalues, then the curve γ is asymptotically stable, i.e., any solution $\varphi(t)$ starting close enough to γ has the property that* $\mathrm{dist}(\varphi(t), \gamma) \to 0$ *as $t \to +\infty$.*

Proposition 2.9. *If the curve γ is hyperbolic, then there exist invariant submanifolds W^s (stable) and W^u (unstable), such that $\mathrm{dist}(g^t(x), \gamma) \to 0$ for $x \in W^s$ as $t \to \infty$ and $\mathrm{dist}(g^t(y), \gamma) \to 0$ for $y \in W^u$ as $t \to -\infty$.*

More interesting is the following proposition.

Proposition 2.10. *If $n = \dim M = 2$, the manifold M and the vector field $v(x)$ are analytic, and γ is a closed phase curve of v, then*

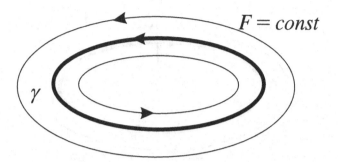

Fig. 2.4.   First integral levels.

either γ is a limit cycle or there exists a single-valued first integral in a neighborhood of the curve γ.

Proof.   In fact, here we have to prove that the periodic solutions of v cannot accumulate on the γ curve. This is equivalent to the fact that the Poincaré return map $f : (\mathbb{R}, 0) \longmapsto (\mathbb{R}, 0)$ has either an isolated fixed point at $z = 0$ or $f(z) \equiv z$. But this follows from the analyticity of the function $f(z) - z$ (assuming that S is analytic) and the standard properties of analytic functions. In the case of $f = id$, all the phase curves in the neighborhood of γ are closed and they are the levels of a first integral of F for the vector field (Fig. 2.4).   □

The latter statement has an analogue for the singular point $x = y = 0$ of an analytic vector field if the linear part of the field has non-real values, i.e.,

$$\dot{x} = \alpha x - \omega y + \cdots, \quad \dot{y} = \omega x + \alpha y + \cdots, \quad \omega \neq 0. \qquad (2.1)$$

We have to go to the polar coordinate system. We will get

$$\dot{r} = \alpha r + r^2 A(r, \varphi), \quad \dot{\varphi} = \omega + r B(r, \varphi), \qquad (2.2)$$

where $A(r, \varphi)$ and $B(r, \varphi)$ are expanded into convergent series in powers of r with coefficients being trigonometric polynomials of φ (Exercise 2.54). The phase curves of this system satisfy the differential equation

$$\frac{\mathrm{d}r}{\mathrm{d}\varphi} = r \frac{\alpha + r A(r, \varphi)}{\omega + r B(r, \varphi)}. \qquad (2.3)$$

Its solutions $r = \psi(\varphi; r_0)$, such that $\psi(0; r_0) = r_0$, define the map

$$f : (\mathbb{R}_+, 0) \longmapsto (\mathbb{R}_+, 0), \quad r_0 \longmapsto \psi(2\pi; r_0),$$

which is analogous to the Poincaré return map. In fact, this is a return transformation for field (2.1) from the half-line $S_+ = \{(x, 0) : x \geq 0\} \simeq (\mathbb{R}_+, 0)$ to itself (see Fig. 2.5). By the convergence of the series representing A and B, the transformation of f is analytic. The fixed points of the diffeomorphism f correspond to the closed phase curve of field (2.1). As in the proof of the previous statement, either $r = 0$ is an isolated fixed point for f, or $f = id$ and then all the phase curves in the neighborhood of $x = y = 0$ are closed.

The Poincaré map from the latter proof expands f into the series

$$f(r) = a_1 r + a_2 r^2 + \cdots. \tag{2.4}$$

It is easy to see that $a_1 = \exp(2\pi\alpha/\omega)$ (Exercise 2.55).

Lemma 2.11. *If $a_1 = 1$ then $a_2 = 0$ and, more generally, if $\ln a_1 = a_2 = \cdots = a_{2k-1} = 0$ then $a_{2k} = 0$.*

This lemma is a consequence of the Poincaré–Dulac theorem (Theorem 3.19 from Chapter 3) and therefore we do not prove it here. The reader can prove it by using some symmetry properties ($\varphi \longmapsto \varphi + \pi$) for the functions A and B in Eq. (2.3). One can also consider the "square root" of the Poincaré map, i.e., the map $g : (\mathbb{R}, 0) \longmapsto (\mathbb{R}, 0)$, with $g'(0) < 0$, of the first return to the horizontal axis; it satisfies $g \circ g = f$ (Exercise 2.56).

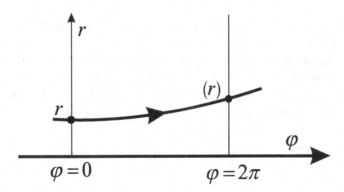

Fig. 2.5.   Return map.

Hence, if $\ln a_1 = a_3 = a_5 = \cdots = a_{2k-1} = 0$ and $a_{2k+1} < 0$ (respectively, > 0) the point $x = y = 0$ is a stable (respectively, unstable) focus.

Definition 2.12. The coefficients

$$c_1 = \frac{1}{2\pi} \ln a_1, \quad c_3 = \frac{1}{2\pi} a_3, \quad c_5 = \frac{1}{2\pi} a_5, \ldots$$

are called the **Poincaré–Lyapunov focus quantities**.

We say that system (2.1) has **weak focus of order** $2n + 1$ if $c_1 = \cdots = c_{2n-1} = 0$ and $c_{2n+1} \neq 0$, if $c_1 \neq 0$ then the focus is **strong**.

Theorem 2.13 (Poincaré–Lyapunov). *In the case of an analytic field of type (2.1), we have following possibilities: either point $(0,0)$ is a focus (stable or unstable) or there exists an analytic single-valued first integral in a neighborhood of this point (i.e., $(0,0)$ is a center).*

Proof. The proof of the existence of a real first integral in the case of vanishing of all focus quantities is the same as in the proof of Proposition 2.10. From the analysis presented below this proof, it follows that there exists a formal power series defining the first integral. It remains to show the analyticity of the first integral.

Here we need an analytic version of the Hadamard–Perron theorem for germs of holomorphic vector fields v in $(\mathbb{C}^2, 0)$. In our case, we have

$$\dot{z} = \lambda z + \cdots, \quad \dot{w} = -\lambda w + \cdots,$$

where $z = x + iy$, $w = x - iy$ and $\lambda = i\omega$; note that, after dividing the right-hand sides by $i = \sqrt{-1}$ (it amounts to changing complex time) we get a hyperbolic singularity. In this case, there exist local analytic invariant manifolds $W^+ = \{z = O\left(w^2\right)\}$ and $W^- = \{w = O\left(z^2\right)\}$.

After passing to local coordinates such that $W^\pm$ are coordinate lines, we get a system $\dot{z} = z(\lambda + \cdots)$, $\dot{w} = w(-\lambda + \cdots)$ with the equation

$$\frac{dw}{dz} = -\frac{w}{z}(1 + \cdots)$$

for the phase curves (which are Riemann surfaces).

Let $D = \{z = z_0, |w| < \varepsilon\}$ be a small disc transversal to $W^- = \{w = 0\}$ at the point $(z, w) = (z_0, 0) \neq (0, 0)$. For any initial value $(z_0, w_0) \in D$ consider the solution $w = \varphi(z; w_0)$ to the latter equations with the initial condition $w(x_0) = w_0$. The solution φ is analytic in the neighborhood of z_0 (by the analytic version of the theorem about the existence and uniqueness of solutions (see Theorem 6.5 from Appendix) and can be continued along paths in the z-plane (the paths cannot pass through $z = 0$). For the path $\{z(\theta) = z_0 e^{i\theta} : 0 \leq \theta \leq 2\pi\}$ around $z = 0$ we have $w(\theta) = \varphi(z(\theta); w_0)$ and $w(2\pi) = \mathcal{M}(w_0)$ defines the monodromy map

$$\mathcal{M} : D \longmapsto D$$

(a complex analogue of the Poincaré map).

Using the Poincaré–Dulac theorem (Theorem 3.19) (in Chapter 3) it can be shown that the vanishing of the focus numbers implies that the latter map is the identity.

Now, we can construct the analytic first integral of the form

$$F(z, w) = zw + \cdots.$$

Firstly, we define it on the section D by $F(z_0, w) = z_0 w$. Then we prolong it above paths in the z-plane avoiding $z = 0$: $F(z, \varphi(z; w_0)) = z_0 w_0$ (constant on the graphs of solutions). So, F can be prolonged to a single-valued function in a neighborhood of $(0, 0)$ deprived the invariant line $W^+ = \{z = 0\}$. It can be expanded into a Laurent series, which cannot have terms with negative powers of z. So, F has removable singularity and can me prolonged to a whole neighborhood of $z = w = 0$. $\qquad \square$

The focus quantities are important when studying the so-called *small amplitude limit cycles*, i.e., those that arise from the focus when the vector field depends on some parameters; for this reason sometimes in the case of order $2n + 1$ we say that the focus has *cyclicity n*.

But these quantities are difficult to compute. Here we present one way to calculate them; this way in fact used by Lyapunov.

Instead of real coordinates, we will use the coordinates $z = x + iy$ and $\bar{z} = x - iy$, $i = \sqrt{-1}$, so that the vector field is written in the form of one complex equation

$$\dot{z} = iz + Az^2 + Bz\bar{z} + C\bar{z}^2 + Dz^3 + Ez^2\bar{z} + Fz\bar{z}^2 + G\bar{z}^3 + \cdots, \quad (2.5)$$

where $A, B, C, D, E, F, G, \ldots$ are complex constants fixed (and we assume $\omega = 1$). Let us note that here the linear part is simpler; in particular, $c_1 = 0$.

We will look for a first integral for the Eq. (2.5) in the form

$$H(z, \bar{z}) = z\bar{z} + a_{30}z^3 + a_{21}z^2\bar{z} + a_{12}z\bar{z}^2 + a_{03}\bar{z}^3 + \cdots, \quad (2.6)$$

where the condition of reality of H leads to the conditions $a_{ji} = \bar{a}_{ij}$. Of course, generally there will be no first integral and the obstacles to this are related to the Poincaré–Lyapunov focus quantities.

The expected property $\dot{H} \equiv 0$ leads to the following set of algebraic equations:

$$(3ia_{30} + \bar{C})z^3 + (ia_{21} + A + \bar{B})z^2\bar{z} \equiv 0$$

for the coefficients in the cubic part of H. We find $a_{30} = i\bar{C}/3$, $a_{21} = i(A + \bar{B})$ and

$$H = z\bar{z} + i\left(\bar{C}z^3 - C\bar{z}^3\right)/3 + i\left((A + \bar{B})z^2\bar{z} - (\bar{A} + B)z\bar{z}^2\right) + \cdots$$

(there are no obstacles here). But for the term at $z^2\bar{z}^2$, after differentiating function (2.6), we get

$$0 \cdot ia_{22} + E + \bar{E} + i(AB - \bar{A}\bar{B}) = 0.$$

We see that, for $\dot{H} = 0$ (modulo of the fifth order), there must be

$$-\operatorname{Im}(AB) + \operatorname{Re}E = 0;$$

we expect that the focus quantity c_3 is proportional to $-\operatorname{Im}(AB) + \operatorname{Re}E$.

To find the constant of proportionality, note that $\dot{\varphi} = 1 + O(r)$, $H = r^2 + O(r^3)$ and $\dot{H} = 2\left(-\operatorname{Im}AB + \operatorname{Re}E\right)r^4 + O(r^5)$. Then

$$f(r) - r = \Delta r \approx \frac{dr}{dH}\Delta H \approx \frac{1}{2r}\int_0^{2\pi}\dot{H}\frac{dt}{d\varphi}d\varphi \approx \frac{1}{2r}$$

$$\cdot 2(-\operatorname{Im}AB + \operatorname{Re}E)r^4 \cdot 1 \cdot 2\pi.$$

This gives the following proposition.

Proposition 2.14. *We have*

$$c_3 = -\operatorname{Im}(AB) + \operatorname{Re}E. \quad (2.7)$$

Remark 2.15. The existence of center is guaranteed by vanishing of infinitely many focus quantities. But sometimes it can be proved directly.

For example, one can look for an explicit formula for the first integral. Often, it is in the so-called Darboux form, like in Theorem 6.21 from Appendix.

Also some singular points, with pure imaginary eigenvalues, of the so-called *time-reversible systems* are centers. The latter systems are invariant with respect to the symmetry $(x, y, t) \longmapsto (-x, y, -t)$, i.e., for some coordinate systems.

Example 2.16 (Quadratic centers and foci). Let us continue calculations of the focus quantities for system (2.5), but only in the quadratic case. i.e., with $D = E = \cdots = 0$. Moreover, assume that $c_3 = 0$, i.e., $\operatorname{Im} AB = 0$ or $A = k\overline{B}$ with $k \in \mathbb{R}$.

Recall that we look for a first integral $H = zw + H_3 + H_4 + \cdots$, where $H_j(z, \bar{z})$ are homogeneous summands of degree j. H_3 is given above and we find

$$H_4 = -\frac{1}{4}\overline{C}\left(2A + \overline{B}\right)z^4 - \frac{1}{2}(2A^2 + 3A\overline{B} - \overline{C}B - 2\overline{C}A + \overline{B}^2)z^3\bar{z} + \text{conj},$$

where conj denotes the conjugate to the previous part. Further calculations give $a_{41} = \frac{i}{3}(-3A^3 - \frac{11}{2}A^2\overline{B} - \frac{1}{2}AB\overline{C} + 3A\overline{CA} - 3A\overline{B}^2 - \frac{1}{2}B\overline{CB} + \overline{CAB} - \frac{1}{2}\overline{B}^3)$, $a_{32} = i(-3A^2B - A^2\overline{A} - \frac{1}{2}\overline{AB}^2 - \frac{9}{2}AB\overline{B} + AC\overline{C} - \frac{3}{2}A\overline{AB} - BC\overline{A} - \frac{3}{2}B\overline{B}^2 + \frac{1}{2}C\overline{CB} - 2\overline{CA}^2)$ and $a_{23} = -i(-2A^2C - \frac{1}{2}AB^2 - 3\overline{A}^2B - \frac{3}{2}AB\overline{A} - \frac{3}{2}B^2\overline{B} - \frac{9}{2}B\overline{AB} + C\overline{CA} + \frac{1}{2}BC\overline{C} - AC\overline{B} - A\overline{A}^2)$.

Finally, we find

$$\dot{H} = (4Ca_{4,1} + 3Ba_{3,2} + 2a_{23} + \text{conj})\, z^2\bar{z}^3 + \cdots$$

$$= \left\{\frac{i}{3}\left(A - 2\overline{B}\right)\left(2A + \overline{B}\right)\overline{B}C - \frac{i}{3}\left(\overline{A} - 2B\right) \right.$$

$$\left. \left(2\overline{A} + B\right)B\overline{C}\right\} z^3\bar{z}^3 + \cdots,$$

which gives (Exercise 2.57)

$$c_5 = -\frac{1}{3}\operatorname{Im}\left[\left(2A + \overline{B}\right)\left(A - 2\overline{B}\right)\overline{B}C\right]. \tag{2.8}$$

Also the next quantity can be found in this way (i.e., assuming $c_3 = c_5 = 0$):

$$c_7 = -\frac{5}{8}(|B|^2 - |C|^2)\operatorname{Im}[(2A + \overline{B})\,\overline{B}^2 C]. \qquad (2.9)$$

It turns out that vanishing of all the above quantities yields that the equilibrium point $z = 0$ becomes a center. In fact, we have four components of the algebraic variety defined by $c_3 = c_5 = c_7 = 0$:

$$
\begin{aligned}
Q_3^H &: 2A + \overline{B} = 0, \\
Q_3^R &: \operatorname{Im} AB = \operatorname{Im} \overline{B}^3 C = \operatorname{Im} A^3 C = 0, \\
Q_3^{LV} &: B = 0, \\
Q_4 &: A - 2\overline{B} = |B|^2 - |C|^2 = 0.
\end{aligned}
\qquad (2.10)
$$

Above the lower index denotes the codimension of the corresponding component in the space of all quadratic systems; thus we must add the condition $c_1 = 0$ (and some inequalities, which we omit).

Using the fact that only the first four focus quantities are sufficient N. Bautin proved that the maximal number of small amplitude limit cycles bifurcating from either a weak focus or a center $z = 0$, i.e., its cyclicity, equals 3.

The case Q_3^H corresponds to Hamiltonian systems, i.e., with zero divergence (Exercise 2.58).

The case Q_3^R correspond to time-reversible systems. Indeed, we can apply a rotation of the variable z such that the coefficients A, B, C become real (Exercise 2.59). Then we get the following system:

$$\dot{x} = -y + ax^2 + by^2, \quad \dot{y} = x + cxy.$$

In the case Q_3^{LV}, we can assume that $A = |A|\,e^{i\theta}$, $B = 0$ and $C \geq 0$ (by rotating z). We claim that our system $\dot{z} = iz + Az^2 + C\bar{z}^2$ has three invariant lines (two of them can be non-real). Indeed, let $e^{i\varphi}z + e^{-i\varphi}\bar{z} = d$ be an equation defining such L. Then we should have

$$e^{i\varphi}\left(iz + Az^2 + C\bar{z}^2\right) + e^{-i\varphi}\left(-i\bar{z} + \overline{A}\bar{z}^2 + Cz^2\right)|_L \equiv 0,$$

which implies $[-i + (Ce^{3i\varphi} + \overline{A}e^{i\varphi})d]\left(d - 2e^{i\varphi}z\right) + [|A|\cos(\varphi - \theta) + C\cos(3\varphi)]\,2e^{2i\varphi}z^2 \equiv 0$. Solving the equation $|A|\cos(\varphi - \theta) + C\cos(3\varphi)$ for φ (with 3 or 2 or one solutions) one finds also d.

Assume that we have three invariant real lines defined by the equations $L_j(x, y) = 0$, $j = 1, 2, 3$, which do not pass through $x = y = 0$. We have $\dot{L}_j = K_j(x, y) L_j$, where the three linear forms $K_j = \alpha_j x + \beta_j y$ are dependent: $\sum \gamma_j K_j \equiv 0$. It follows that there exist constants γ_j such that the function

$$F = L_1^{\gamma_1} L_2^{\gamma_2} L_3^{\gamma_3} \tag{2.11}$$

is a Darboux type first integral (Exercise 2.60). Such systems are special (integrable) cases of Lotka–Volterra systems considered in Example 2.38.

In the case of non-real lines, say $L_2 = \overline{L}_3$ the exponents $\gamma_2 = \bar{\gamma}_3$ are also non-real.

In the codimension 4 case Q_4, we can assume that the system is of the form

$$\dot{z} = iz + 2z^2 + |z|^2 + e^{i\xi}\bar{z}^2. \tag{2.12}$$

It turns out that it has two invariant curves

$$f_2 = x_1^2 + 4y + 1 = 0, \quad f_3 = \cos(\xi/2) x_1 \left(x_1^2 + 6y\right) + 6y + 1 = 0, \tag{2.13}$$

where $x_1 = 2\,\mathrm{Im}\left(e^{-i\xi}z\right)$ and we have $\dot{f}_j = 2jx f_j$ (Exercise 2.61). Thus, we have the first integral

$$F = f_2^3/f_3^2. \tag{2.14}$$

One can say more about the above focus numbers c_3, c_5 and c_7. Of course, they are homogeneous polynomials of degree 2, 4 and 6, respectively.

But they are also quasi-homogeneous of degree 0 with respect to the following grading (or weights):

$$w(A) = w(\overline{B}) = -w(\overline{A}) = -w(B) = 1, \quad w(C) = -w(\overline{C}) = -3.$$

This is related with the changes they undergo when we apply the rotations $z \mapsto e^{i\theta}z$, $\bar{z} \mapsto e^{-i\theta}\bar{z}$.

Using this and the fact that c_j vanish when the center conditions (2.10) are satisfied, allows to determine the forms (2.7)–(2.9) focus quantities up to multiplicative constants. Next, those constants are calculated after perturbations of concrete cases of center.[b]

---

[b]For more information we refer the reader to a work in H. Żołądek, Quadratic systems with center and their perturbations, *J. Differential Equations* 109 (1994), 223–273, by the first author.

Example 2.17 (Focus quantities for Liénard equation). The standard **Liénard equation**

$$\ddot{x} + f(x)\dot{x} + g(x) = 0 \qquad (2.15)$$

is often rewritten as the following **Liénard system:**

$$\dot{x} = y, \quad \dot{y} = -g(x) - f(x)y. \qquad (2.16)$$

Here we assume that the functions appearing in the right-hand sides are analytic near $x = 0$ and have the following Taylor expansions:

$$f(x) = \sum_{n \geq 1} a_n x^n, \quad g(x) = \sum_{n \geq 1} b_n x^n.$$

Assume also that

$$b_1 > 0; \qquad (2.17)$$

so, the linear part at $(0,0)$ has the imaginary eigenvalues $\pm i\sqrt{b_1}$. We introduce the primitives of the above functions

$$F(x) = \int_0^x f(s)\,\mathrm{d}s, \quad G(x) = \int_0^x g(s)\,\mathrm{d}s.$$

Under assumption (2.17), we can make a change $x \to u = \sqrt{b_1}x + \cdots$ such that

$$G(x) = u^2/2;$$

thus $u = \sqrt{2G(x)}$ and $\frac{\mathrm{d}u}{\mathrm{d}x} = \frac{g(x)}{u}$. Then system (2.16) becomes $\dot{u} = \frac{g(x)}{u}y$, $\dot{y} = \frac{g(x)}{u}\{-\frac{f(x)}{g(x)}u \cdot y - u\}$. Let us multiply the right-hand sides by $\frac{u}{g(x)}$; it corresponds to introduction of a new time τ (but the phase portrait becomes unchanged). We arrive at the simplified Liénard system

$$\frac{\mathrm{d}u}{\mathrm{d}\tau} = y, \quad \frac{\mathrm{d}y}{\mathrm{d}\tau} = -u - k(u)y, \qquad (2.18)$$

where

$$k(u) = uf(x)/g(x) = (a_1/\sqrt{b_1})u + \cdots.$$

Note that

$$K(u) = \int_0^u k(z)\,\mathrm{d}z = \sum_{j \geq 2} k_j u^j = \int_0^x \frac{\mathrm{d}s}{\mathrm{d}z}f(s)\,\mathrm{d}s = F(x).$$

If $K(u) \equiv 0$ then system (2.18) is Hamiltonian with the first integral $H_0 = \frac{1}{2}u^2 + \frac{1}{2}y^2$, i.e., $(0,0)$ is a center. Also when the function

K is even, i.e., $K(u) = \widetilde{K}(u^2)$, then $(0,0)$ is a center, because then system (2.18) is reversible with respect to the involution $(u, y, \tau) \mapsto (-u, y, -\tau)$. In fact, we have a first integral

$$H(u, y) = \widetilde{H}(u^2, y) = H_0 + \cdots ;$$

we do not know an explicit formula for it, but we do not need it for our aims.

Suppose now that the series $K(u)$ is not even, i.e., $k_3 = k_5 = \cdots = k_{2n-1} = 0$, but $k_{2n+1} \neq 0$. Thus, the polynomial system $\frac{du}{d\tau} = y$, $\frac{dy}{d\tau} = -u - \left(\sum_{j \leq n} 2jk_{2j} u^{2j-1} \right) y$ has a first integral H (as above), but the whole system is not such.

Like in the proof of Proposition 2.14 we compute approximately the Poincaré return map by estimating the increment of the function H. We get

$$\Delta H = \int_0^T \frac{dH}{d\tau} d\tau = -\int_0^T H_y' \cdot \left\{ (2n+1) k_{2n+1} u^{2n} + \cdots \right\} y \cdot d\tau$$

$$\approx -(2n+1) k_{2n+1} \int y \cdot u^{2n} \cdot du \approx -(2n+1) k_{2n+1} \iint u^{2n} dy du$$

$$\approx \text{const} \cdot k_{2n+1} h^{n+1};$$

above we replaced $y d\tau$ with du and a piece of a phase curve with its approximation $\{H = h\}$. This implies that

$$c_3 = c_5 = \cdots = c_{2n-1} = 0, \quad c_{2n+1} = \text{const} \cdot k_{2n+1} \neq 0, \quad (2.19)$$

i.e., we have weak focus of order $2n + 1$.

Finally, we note the following interpretation of this order. Consider two analytic plane curves $u_1^2 = u_2^2$ and $K(u_1) = K(u_2)$. Of course, they have the common component $u_1 = u_2$. The other components are

$$u_1 + u_2 = 0 \text{ and } k_2(u_1 + u_2) + k_3(u_1^2 + u_1 u_2 + u_2^2) + \cdots = 0.$$

If $K(u)$ is even, then the latter two curves coincide; otherwise, they are tangent at $u_1 = u_2 = 0$ with the order of tangency equal $2n$; the second curve is of the form $u_2 = -u_1 + \text{const} \cdot u_1^{2n} + \cdots$ (Exercise 2.62).

The same holds for the curves

$$G(x_1) = G(x_2) \quad \text{and} \quad F(x_1) = F(x_2).\tag{2.20}$$

The order of the weak focus $(0,0)$ of the Liénard system (2.16) equals 1 plus the tangency order at $x_1 = x_2 = 0$ of the non-obvious components of the analytic plane curves (2.20).

2.2. Poincaré–Bendixson criterion

The problem of detecting limit cycles turns out to be very difficult. This is demonstrated by the following unresolved problem.

Conjecture 2.18 (Hilbert's 16th problem).[c] *Give an esti-mate in terms of degrees of polynomials P and Q for the number of polynomial cycles of the vector field of form*

$$\dot{x} = P(x,y), \quad \dot{y} = Q(x,y).\tag{2.21}$$

Remark 2.19. It is known that the number of limit cycles of an indi-vidual system of the form (2.21) is finite (Yu. Ilyashenko, J. Ecalle), but it is not known whether there exists a bound for this number in terms of $n = \max(\deg P, \deg Q)$ (Exercise 2.63). There exist exam-ples of quadratic systems with four limit cycles, but even in this case it is not known where such an upper bound exists.

Positive solution was obtained for the so-called weakened 16th Hilbert problem for the number of zeroes of the following Abelian integral:

$$I(h) = \oint \omega\tag{2.22}$$

of a polynomial 1-form $\omega = A(x,y)\,dx + B(x,y)\,dy$ over an oval $\gamma_h \subset \{H = h\}$, i.e., compact component diffeomorphic to a circle; such integrals (called also Melnikov functions) arise after polynomial perturbations of Hamiltonian systems with the polynomial Hamilton

---

[c]In essence, this is the second part of the 16th Hilbert problem. The first part deals with the number and position of connected components (the so-called ovals) of real algebraic curves of the form $F(x,y) = 0$. Here the problem is largely solved (with appropriate generalizations).

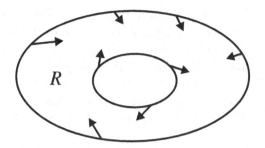

Fig. 2.6. Absorbing ring.

function $H(x, y)$ (see also Section 4.1.3). There exists a double exponential estimate for the number of such zeroes, due to G. Binyamini, D. Novikov and A. Yakovenko.

Therefore, concrete methods showing the existence of limit cycles are important. The following Poincaré–Bendixson criterion guarantees us the existence of at least one limit cycle provided the vector field is analytic (see Proposition 2.10).

Let us assume that we have a vector field $v(x)$ on the plane and a domain $\mathcal{R} \subset \mathbb{R}^2$ of ring type (like in Fig. 2.6) with the following properties:

- the field $v(x)$ has no equilibrium points in $\mathcal{R}$,
- the field $v(x)$ on the boundary $\partial \mathcal{R}$ of the ring $\mathcal{R}$ is directed into its interior.

Theorem 2.20 (Poincaré–Bendixson). *Under these assumptions, there is at least one closed phase curve of the field v inside $\mathcal{R}$.*

The proof of this theorem is based on the following important concept in Dynamic Systems Theory.

Definition 2.21. The ω-**limit set of the point** x, denoted $\omega(x)$, for a phase flow $\{g^t\}$ (or a cascade $\{f^n\}$) is the set of accumulation points of the positive orbit of this point, i.e.,

$$\omega(x) = \{y : \exists t_k \to +\infty \text{ such that } g^{t_k}(x) \to y\}$$

(or $\omega(x) = \{y : \exists n_k \to \infty \text{ such that } f^{n_k}(x) \to y\}$) (Exercise 2.64).

In the case of the accumulation points of the negative orbit of the point x (i.e., when $t_k \to -\infty$ or $n_k \to -\infty$), it is called the α-**limit set of the point** x.

Obviously, an attracting equilibrium point (or an attracting limit cycle) is the ω-limit set of any point lying close to this equilibrium (or this cycle). More complex ω-limit set can consists of several saddles connected by separatrices, the so-called polycycle. Also the ω-limit set can be empty, e.g., when the trajectory escapes to infinity.

There is a version of Poincaré–Bendixson theorem which uses the notions of the ω-limit set for the phase flow generated by a vector field.

Proposition 2.22 (Poincaré–Bendixson). *If, for a vector field v in $\mathbb{R}^2$ and a point x, the set $\omega(x)$ is:*

(a) *bounded,*
(b) *does not contain equilibrium points of the field,*

then $\omega(x)$ is a closed phase curve of this field.

Proof. Let $y \in \omega(x)$ and let $\Gamma(y) = \{g_v^t(y) : t \in \mathbb{R}\}$ be the trajectory of the field passing through y. Of course, $\Gamma(y) \subset \omega(x)$; hence, also $\omega(y) \subset \omega(x)$ (as $\omega(x)$ is closed). We will show that the trajectory $\Gamma(y)$ is closed; in such case it must be a closed phase curve (due to the assumptions (a) and (b)) .

To do this, let us choose a local section S perpendicular to $v(y) \neq 0$ at y. Consider the intersection points $x_k = g_v^{t_k}(x)$, $t_{k+1} > t_k$, of the positive orbit $\Gamma_+(x) = \{g_v^t(x) : t \geq 0\}$ with the section S. By assumption, there are infinitely many such points and we can assume that the sequence $\{x_k\}$ is monotonous on S; here we use the fact that we are on the plane (Exercise 2.65). So, we have the situation as in Fig. 2.7.

Let us assume that the trajectory $\Gamma(y)$ is not a closed phase curve. Then $\omega(y) \neq \Gamma(y)$ and there exists an accumulation point $z \in \omega(y) \backslash \Gamma(y)$ of the trajectory $\Gamma(y)$. Again, we can take a section S_1 perpendicular to $v(z)$ at z and we obtain the situation like at Fig. 2.8 (possibly replacing y with some of the points y_k in the intersection of $\Gamma(y)$ with S_1).

Now, by deforming the part of the trajectory $\Gamma(y)$ (from y to y_1) such that the new curve is positioned at a non-zero (but small) angle to the field v, we get a domain $\Omega \subset \mathbb{R}^2$ to which the field is "entering"; the domain Ω can be bounded or unbounded. But this leads to a

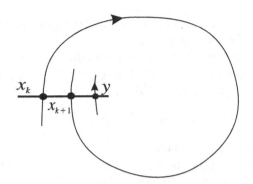

Fig. 2.7. Successive returns.

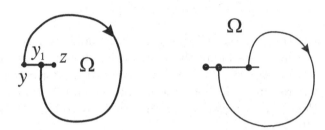

Fig. 2.8. Invariant domain Ω.

contradiction, because then it must be $\Gamma(y_1) \subset \Omega$, and hence also

$$\omega(x) \subset \Omega.$$

Note that $\omega(x)$ must also contain the points of the negative orbit $\Gamma_-(y) = \{g^t(y) : t < 0\}$ (of the point y) outside Ω. $\qquad\square$

Proof of Theorem 2.20. We take any point $x \in \partial\mathcal{R}$. Then its ω–limit set satisfies the assumptions of Proposition 2.22.

Example 2.23. Let us consider the following case of the so-called **Liénard system**[d]

$$\dot{x} = y, \quad \dot{y} = -x - y + F(y), \qquad (2.23)$$

---

[d]This example comes from the monograph *Modern Geometry — Methods and Applications* by B. A. Dubrovin, A. T. Fomenko and S. P. Novikov, Springer (1985). Unfortunately, some significant details are absent there. We have completed them.

Moreover, the Liénard system usually assumes either of the form $\dot{x} = y$, $\dot{y} = -f(x)y - g(x)$ (i.e., Eq. (2.16) or of the form $\dot{x} = y + F(x)$, $\dot{y} = -g(x)$). System (2.23), after swapping x with y, is reduced to the second one.

where $F(y) = 2y/\sqrt{1 + 4y^2}$; in fact, the point is that F is odd, $F'(0) > 1$ and $|F(\pm\infty)|$ is finite.

Let us note that the only point of equilibrium $x = y = 0$ is unstable focus (with the eigenvalues $\frac{1}{2}(1 \pm i\sqrt{3})$). Therefore, we choose the inner boundary of the ring $\mathcal{R}$ (to apply Theorem 2.20) in the form

$$\partial_{\text{in}}\mathcal{R} = \left\{ x^2 + y^2 = \rho^2 \right\}$$

for small $\rho > 0$ (Exercise 2.66).

We would like to take the outer border in the form of a large circle $x^2 + y^2 = R^2$. Unfortunately, the identity

$$\frac{\mathrm{d}}{\mathrm{d}t} \left(x^2 + y^2 \right) = 2y \left(F(y) - y \right)$$

shows that in the domain $\{F(y)/y > 1\}$ the "radius square" function $r^2 = x^2 + y^2$ increases along the trajectories. Fortunately, this bad domain is small. Anyway, one has to perform some estimates.[e]

As a preparation to the construction of the outer boundary of the Poincaré–Bendixson ring we divide the phase space into four subsets (see Fig. 2.9):

$$U_{I,III} = \left\{ \pm y \geq 1/\sqrt{\varepsilon} \right\}, \quad U_{II,IV} = \left\{ |y| \leq 1/\sqrt{\varepsilon}, \pm x \geq 0 \right\}.$$

Here $\varepsilon > 0$ is a small constant such that the function $F(y)$ is close to ± 1 in $U_{I,III}$, we have $|F(y) \mp 1| < \varepsilon$. Therefore, in $U_{I,III}$ the equation for phase curves of (2.23) in the standard polar coordinates is

$$\frac{\mathrm{d}r}{\mathrm{d}\varphi} = \frac{y(y - F(y))}{r \left(1 + \sin(2\varphi)/2 - xF(y)/r^2 \right)} \approx \frac{y^2}{r(1 + \sin(2\varphi)/2)}. \quad (2.24)$$

In the domains $U_{II,IV}$, when $|x|$ is large ($|x| > 1/\varepsilon \gg |y|$), the phase curves satisfy

$$\frac{\mathrm{d}x}{\mathrm{d}y} = \frac{-y}{x + y - F(y)} \approx \frac{-y}{x} \quad (2.25)$$

and are approximately vertical.

---

[e]This situation is rather usual in the Qualitative Theory. Although general statements are geometric and quite clear, their application in real systems requires estimates which can be technical and complicated.

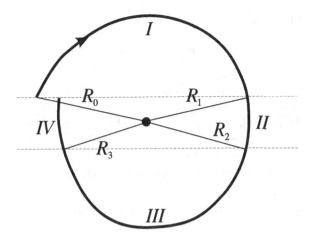

Fig. 2.9. Liénard system.

Take R_0 large ($R_0 \gg 1/\varepsilon$) and consider the phase curve Γ which starts at a point $(x_0, y_0) \in U_I \cap U_{IV}$ such that the radius $r = R_0$ (Fig. 2.9). We consider the pieces $\Gamma_I, \ldots, \Gamma_{IV}$ of Γ in the corresponding domains. If Γ_{IV} ended at a radius R_4 smaller than R_1 then the outer boundary $\partial_{\text{out}} \mathcal{R}$ could be chosen as a slight modification of the curve consisting of $\Gamma_I \cup \cdots \cup \Gamma_{IV}$ and a segment joining the suitable endpoints of Γ_I and Γ_{IV}.

For Γ_I in Eq. (2.24), we have $0 < \varphi_1 \le \varphi \le \varphi_2 < \pi$, where $\varphi_1 \to 0$ and $\varphi_2 \to \pi$ as $R_0 \sqrt{\varepsilon}$ and $R_1 \sqrt{\varepsilon}$ tend to infinity. Moreover, we have the estimates

$$C_1 r \sin^2 \varphi < \frac{\mathrm{d}r}{\mathrm{d}\varphi} < C_2 r \sin^2 \varphi$$

for constants $C_1 \approx 1$ and $C_2 \approx 2/3$; recall also the boundary conditions $r(\varphi_2) = R_0$ and $r(\varphi_1) = R_1$, where R_1 is the radius value at $\Gamma_I \cap \Gamma_{II}$. We get the inequality

$$a_1 R_0 < R_1 < a_2 R_0, \tag{2.26}$$

where $0 < a_1 < a_2 < 1$ are constants, $a_{1,2} = \exp\left\{-C_{2,1} \int \sin^2 \varphi \mathrm{d}\varphi\right\}$ (Exercise 2.66). Of course, $R_1 < R_0$, but it is still large.

In the domain U_{II}, we have Eq. (2.25), which implies that $|\mathrm{d}y/\mathrm{d}x|$ is small and Γ_{II} is almost vertical. It follows that the radius value

R_2 at $\Gamma_{II} \cap \Gamma_{III}$ satisfies

$$|R_2 - R_1| < b \qquad (2.27)$$

(for a constant b not depending on R_0 and ε) and is still large (Exercise 2.66).

The analysis in the domains U_{III} and U_{IV} is analogous as in the domains U_I and U_{II}, we have

$$a_1 R_2 < R_3 < a_2 R_2, \quad |R_4 - R_3| < b.$$

Summing up the above inequalities, we find that

$$R_4 < R_0.$$

Example 2.24 (FitzHugh–Nagumo model). This model describes dynamics of a neuron and takes form of the following system:

$$\dot{x} = a + x - y - x^3/3, \quad \dot{y} = b + cx - dy, \qquad (2.28)$$

where a, b, c, d are positive constants; it can be reduced to some Liénard system (2.16).

The equations for the singular points, i.e., $f(x) = b - ad + (c - d)x + dx^3/3$ and $y = a + x - x^3/3$, have one or two or three solutions. Assume that there is only one solution (x_0, y_0). We can replace the parameter b with another parameter x_0, i.e., we put

$$b = ad + (d - c)x_0 - dx_0^3/3$$

and get

$$f(x) = (x - x_0)\left(c - d + dx_0^2/3 + dx_0 x/3 + dx^2/3\right) = (x - x_0)g(x).$$

So, we have the condition

$$4c - 4d + dx_0^2 > 0, \qquad (2.29)$$

i.e., the negativity of the discriminant of the quadratic polynomial $g(x)$.

The characteristic polynomial of the linearization matrix A_0 at (x_0, y_0) equals

$$P(\lambda) = \lambda^2 + \left(x_0^2 + d - 1\right)\lambda + c - d + dx_0^2.$$

Assume additionally that

$$c - d + dx_0^2 > 0, \quad x_0^2 + d - 1 < 0;$$

in fact, the first follows from Eq. (2.29). Thus, (x_0, y_0) is either an unstable node or an unstable focus.

The imposed conditions are compatible with the assumption $a, b, c, d > 0$. Indeed, they are satisfied for $x_0 = -\frac{1}{2}$, $a = 1$, $c = d = \frac{1}{4}$ and $b = d\left(a - x_0^3/3\right) = \frac{25}{96}$.

We claim that, under the above restrictions on the parameters, the FitzHugh–Nagumo system has at least one limit cycle.

For this we shall construct a ring $\mathcal{R}$ satisfying the assumptions of the Poincaré–Bendixson theorem. Of course, the interior boundary of $\mathcal{R}$ is a small ellipse around (x_0, y_0).

The exterior boundary is taken in the form

$$L(x, y) := x^4 + y^2/2d = M$$

for some large constant M. We have

$$\dot{L} = -4x^6/3 - 4x^3 y - y^2 + \cdots,$$

where the dots denote lower order terms with respect to the so-called quasi-homogeneous grading such that

$$\deg x = 1, \quad \deg y = 3.$$

Since the discriminant of the quadratic polynomial in x^3 and y on the right-hand side of $\dot{L}$ is negative, this form is negative definite and $\dot{L}$ negative for large $|x|^3 + |y|$ (Exercise 2.67). In particular, $\dot{L}|_{L=M} < 0$ for large M.

2.3. Index

The readers may ask why, in the Poincaré–Bendixson theorem, the domain $\mathcal{R}$ is a ring; maybe it would be enough to be bounded and

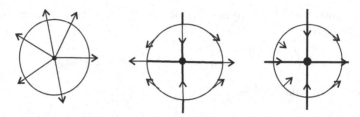

Fig. 2.10.   Index.

simply-connected (i.e., without a hole in the middle). Well, no, and the reason lies in the following theorem.

Theorem 2.25. *Inside the domain bounded by a closed phase curve of a vector field in the plane there is at least one singular point of this field.*

The proof of this result uses topological methods, more specifically, the concept of index.

Definition 2.26. Let $v(x)$ be a vector field in $\mathbb{R}^2$ and let $C \subset \mathbb{R}^2$ be an oriented curve such that

$$v|_C \neq 0. \tag{2.30}$$

The **index $i_C v$ of the field v along the curve C** is the number of rotations of the vector $v|_C$.

If x_0 is an isolated singular point of field v, then the **index $i_{x_0} v$ of the field v at x_0** is the index $i_{C(x_0,\varepsilon)} v$ of the field v along the circle $C(x_0, \varepsilon)$ around x_0 with sufficiently small radius ε and counterclockwise (i.e., positively) oriented.[f]

Example 2.27. For the field $v = x\partial_x + y\partial_y$ we have $i_{(0,0)} v = 1$, for the field $v = x\partial_x - y\partial_y$, we have $i_{(0,0)} v = -1$ and for the field $v = x^2\partial_x - y\partial_y$ we have $i_{(0,0)} v = 0$, see Fig. 2.10 (Exercises 2.68–2.70).

---

[f]An index of an isolated singular point x_0 of the vector field $v(x)$ is generalized to the case of $\mathbb{R}^n$. This is the degree of mapping $x \longmapsto v(x)/|v(x)|$ from a small sphere around x_0 to $\mathbb{S}^{n-1}$.

Proposition 2.28. *For the positively oriented curve C, we have*

$$i_C v = \sum i_{x_j} v,$$

where the summation runs over the singular points within a region bounded by the curve C.

Proof. Let us observe that the mapping

$$(v, C) \longmapsto i_C v$$

is a continuous function on the space of pairs (vector field, curve) satisfying condition (2.30). Since the set of values of this function consists of integers, the index is locally constant. In particular, it does not depend on the deformation of the field and the deformation of the curve (in the class (2.30)); this justifies the definition of the index at a point.

We can deform our curve to a curve C' which is composed of small loops around the equilibrium points x_j and a system of segments that join these points to some base point. Since the field rotations along the segments cancel pairwise, $i_{C_1} v$ is equal to the sum of the field rotations around the singular points (see Fig. 2.11). □

Lemma 2.29. *Let C be a closed phase curve of the field v. If C is positively oriented then $i_C v = 1$ and otherwise $i_C v = -1$.*

Proof. By using the invariance of the index with respect to deformation (in class (2.30)) we can deform the curve and the field to get

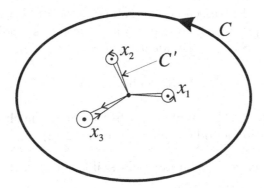

Fig. 2.11. Homotopy in action.

$C = \{x^2 + y^2 = 1\}$ (with positive or negative orientation) and v will be tangent to that curve. It is easy to see that the angle of the vector $v(x, y)$ to C is "delayed" or "accelerated" in relation to the angle of the point (x, y) by $\pi/2$. □

The following corollary implies Theorem 2.25.

Corollary 2.30. *If $\gamma \subset \mathbb{R}^2$ is a closed phase curve of $v(x)$, then*

$$\sum i_{x_j} v = 1,$$

where the summation runs over the singular points x_j of the field within the domain enclosed by the γ curve.

Proof. If γ is positively oriented then we apply Proposition 2.28. Otherwise, we reverse the orientation of $\gamma : C = -\gamma$. □

Corollary 2.31. *Suppose that x_0 is a degenerate singular point of a vector field v, $\det \partial v / \partial x(x_0) = 0$, and let $v_\varepsilon(x) = v(x) + \varepsilon v_1(x; \varepsilon)$ be a deformation of v, such that it has only non-degenerate singular points x_j, $j = 1, \ldots, k$, in a neighborhood of x_0. Then $i_{x_0} v = \sum i_{x_j} v_\varepsilon$.*

Proof. It follows from Proposition 2.28. □

2.4. Dulac criterion

Let us consider a vector field $v(x)$, $x \in \mathbb{R}^2$, with a closed phase curve γ, i.e., the periodic trajectory $x = \varphi_0(t)$ of period T. Let us consider a section S (perpendicular to γ at x_0) and the corresponding Poincaré return map $f : S \longmapsto S$. By identifying S with $(\mathbb{R}, 0)$, we have

$$f(z) = \lambda z + O(z^2), \quad \lambda > 0.$$

Definition 2.32. The number $\mu = \ln \lambda$ is called the **characteristic exponent** of the periodic orbit γ (Exercise 2.72).

If $\mu < 0$ then γ is asymptotically stable (or attracting), whereas if $\mu > 0$ then γ is repelling.

Theorem 2.33 (Dulac). *We have*

$$\mu = \int_0^T \operatorname{div} v(\varphi_0(t))dt,$$

where

$$\operatorname{div} v(x) = \partial v_1/\partial x_1 + \partial v_2/\partial x_2$$

denotes the divergence of the vector field $v(x) = (v_1(x), v_2(x)) = v_1(x)\frac{\partial}{\partial x_1} + v_2(x)\frac{\partial}{\partial x_2}$ (see Appendix).

Proof. Let us consider the equation in variations with respect to the initial conditions for the solution $\varphi_0(t)$ (with the initial condition $\varphi_0(0) = x_0$),

$$\dot{y} = A(t)y, \qquad A(t) = \frac{\partial v}{\partial x}(\varphi_0(t))$$

(see Appendix). Choose two initial conditions for this equation: $y(0) = y_1$, as a unit vector tangent to γ at x_0, and $y(0) = y_2$ as a unit vector perpendicular to γ at x_0. They correspond to two types of perturbations of the initial condition for the equation $\dot{x} = v(x)$: $x(0) = x_0 + \varepsilon y_1$ and $x(0) = x_0 + \varepsilon y_2$.

The solutions $y = \psi_1(t)$ and $y = \psi_2(t)$, satisfying the above initial conditions, span a parallelogram whose area is equal to the Wronskian $W(t) = \det(\psi_1(t), \psi_2(t))$ (see Appendix). For $t = T$ we have $\psi_1(T) = \psi_1(0) = y_1$, because $x = \varphi_0(t) + \varepsilon\psi_1(t) + O(\varepsilon^2)$ actually represents the periodic solution $\varphi_0(t)$ with a slightly shifted starting point (at γ). Next, $\psi_2(T)$ is a vector associated with the solution $x = \varphi_0(t) + \varepsilon\psi_2(t) + O(\varepsilon^2)$ for $t = T$, which does not even need to hit S (see Fig. 2.12). But the projection $\varepsilon(\psi_2(T), y_2)y_2$ of the vector $\varepsilon\psi_2(T)$ onto S has a natural interpretation of $f(\varepsilon y_1) + O(\varepsilon^2)$, where f is the return map.

Let us now note that the parallelogram spanned by the vectors $\psi_1(T) = y_1$ and $\psi_2(T)$ has the same area $W(T)$ as the length of the projection of the vector $\psi_2(T)$ onto S. This means that

$$f'(0) = W(T).$$

By the Liouville theorem (Theorem 6.24 from Appendix) we have $\dot{W} = \operatorname{tr}A(t) \cdot W$ with $\operatorname{tr}A(t) = \operatorname{div}v(\varphi_0(t))$, and this allows us to calculate W. Since $W(0) = 1$, we have $W(T) = \exp\int_0^T \operatorname{tr}A(t)dt$. $\qquad\square$

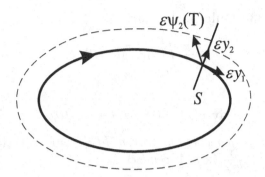

Fig. 2.12.   Dulac theorem.

Dulac theorem turns out to be useful in showing the absence of limit cycles for certain vector fields. We illustrate this in the following example.

Example 2.34 (Jouanolou system). It has the following form:

$$\dot{x} = y^s - x^{s+1}, \quad \dot{y} = 1 - yx^s.$$

According to Theorem 2.25, every limit cycle of this field should surround at least one singular point. Equations of singular points, i.e., $y = x^{-s}$ and $(x^{-s})^s = x^{s+1}$, lead to $x^{s^2+s+1} = 1$. So, there is only one (real) equilibrium point $x = y = 1$.

The linear part of the system at this point is given by the matrix

$$\begin{pmatrix} -s-1 & s \\ -s & -1 \end{pmatrix}$$

with the characteristic equation $\lambda^2 + (s+2)\lambda + (s^2 + s + 1) = 0$. So, the point $(1,1)$ is a stable focus.

Let us suppose that γ is the limit cycle around $(1,1)$ and closest to this point (all limit cycles form a "nest" around $(1,1)$). It is easy to see that γ must be unstable (at least from the inside).

On the other hand, the divergence of the Jouanolou field is

$$\operatorname{div} v = -(s+2)x^s.$$

We can see that, if s is even, then $\operatorname{div} v < 0$ (for almost all points of the phase curve) and Dulac theorem implies that the characteristic exponent of γ is negative (in contradiction with the instability of γ).

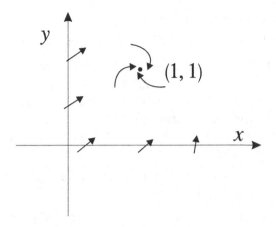

Fig. 2.13.   Jouanolou system.

Note also that the value of the divergence at a singular point is the same as the trace of the corresponding linearization matrix.

If s is odd then this argument also works, but first we need to show that γ must lie in the first quadrant, see Fig. 2.13 (Exercise 2.73).

Dulac theorem has other applications.

Definition 2.35. A function $\Phi : \Omega \longmapsto \mathbb{R}_+$, $\Omega \subset \mathbb{R}^2$, is called the **Dulac function** for a vector field v, if $\operatorname{div}(\Phi v)$ has a fixed sign in the domain Ω.

Theorem 2.36. *If there exists a Dulac function Φ in a region Ω for a vector field v, then every limit cycle lying in Ω is stable when $\operatorname{div}(\Phi v) < 0$ (respectively, unstable when $\operatorname{div}(\Phi v) > 0$).*

Proof. Multiplying a vector field by positive functions does not change the phase portrait of this field. Only the speed of a point changes along the phase curve. $\qquad\square$

Remark 2.37. In Theorem 2.36, the inequalities $\operatorname{div}(\Phi v) \leq 0$ or $\operatorname{div}(\Phi v) \geq 0$ can be admitted. But then we have to exclude the possibility that a possible cycle is completely included in the curve $\{\operatorname{div}(\Phi v) = 0\}$.

Example 2.38 (Generalized Lotka–Volterra system). This is a system describing the evolution of populations of predators and preys (like wolves and rabbits):

$$\dot{x} = x\left[Ay - B(1 - x - y)\right], \quad \dot{y} = y\left[C(1 - x - y) - Dx\right]$$

(with $ABCD \neq 0$) in the domain $x, y > 0$. The equations $Ay = B(1 - x - y)$ and $C(1 - x - y) = Dx$ define the singular point (x_0, y_0), about which we assume that it lies in the first quadrant and that the determinant of the linearization matrix at (x_0, y_0) is positive (only then the index of the field in (x_0, y_0) is 1 and there is a chance for a limit cycle). We claim that:

if $A = D$ then we have a center in (x_0, y_0) and if $A \neq D$ there are no periodic solutions.

To confirm this, consider the following candidate for the Dulac function:

$$\Phi = x^{C/A-1}y^{B/D-1}.$$

We check (using $z = 1 - x - y$):

$$\text{div}\,(\Phi v) = \frac{\partial}{\partial x}x^{C/A}y^{B/D-1}\left(Ay - Bz\right) + \frac{\partial}{\partial y}x^{C/A-1}y^{B/D}\left(Cz - Dx\right)$$

$$= x^{C/A-1}y^{B/D-1}\left\{\frac{C}{A}\left(Ay - Bz\right) + Bx + \frac{B}{D}\left(Cz - Dx\right) - Cy\right\}$$

$$= \frac{BC}{AD}(A - D)x^{C/A-1}y^{B/D-1}z.$$

If $A = D$ then the field Φv is Hamiltonian and has the first integral (Fig. 2.14).

If $A - D \neq 0$, then Φ is the Dulac function in the domain $z > 0$ or in the domain $z < 0$. But the identity

$$\dot{z}|_{z=0} = (D - A)x(1 - x) \neq 0$$

(for $A - D \neq 0$) implies that the possible limit cycle cannot pass through the segment $\{x + y = 1, x, y > 0\}$. Further reasoning is the same as in the Jouanolou example.

Example 2.39 (van der Pol equation). This is the following system:

$$\dot{x} = y, \quad \dot{y} = -x - a(x^2 - 1)y, \quad a > 0,$$

special case of the Liénard system. It appears in electrical engineering, for a system consisting of a capacitor of capacity C, a coil of

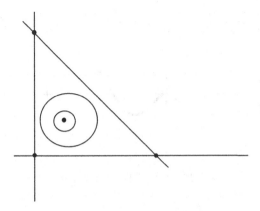

Fig. 2.14.   Lotka–Volterra system.

inductance L and a certain nonlinear element (a diode) replacing a resistor. In case of LCR system (coil, capacitor, resistor), the equation of potential differences gives the equation $L\ddot{I} + R\dot{I} + I/C = 0$ for the current of I in the circuit; in our case the term $\frac{d}{dt}(RI)$ is replaced by the term $\frac{d}{dt}\left[R\left(I^3/3 - I\right)\right]$, i.e.,

$$L\ddot{I} + R(I^2 - 1)\dot{I} + I/C = 0;$$

it is the original **van der Pol equation**. After replacing $t \longmapsto \alpha t = \sqrt{CL}t$ and substituting $x = I$ and $y = \dot{I}$, we obtain the above van der Pol system with $a = R/\alpha$. For more details, we refer the reader to D. Arrowsmith and C. Place's book [7].

We will prove that:

The van der Pol system has exactly one limit cycle which is hyperbolic stable.

Firstly, we note that $(0, 0)$ is the only singular point with the linearization matrix $\begin{pmatrix} 0 & 1 \\ -1 & a \end{pmatrix}$, that is, $\operatorname{Re} \lambda_{1,2} > 0$ and this point is unstable (focus or node).

Let us consider the function

$$\begin{aligned}
f(x, y) &= y^2 + a(x^3/3 - x)y + x^2 - c \\
&= \left[y + \tfrac{1}{2}a\left(x^3/3 - x\right)\right]^2 - g(x^2) - c \\
&= y_1^2 - g(x^2) - c,
\end{aligned}$$

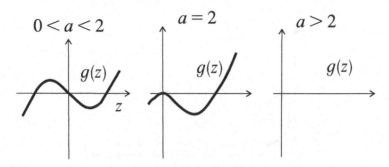

Fig. 2.15. Function g.

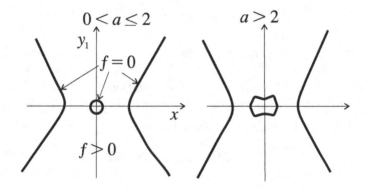

Fig. 2.16. Function f.

where

$$g(z) = \frac{a^2}{36}z^3 - \frac{a^2}{6}z^2 + \left(\frac{a^2}{4} - 1\right)z$$

and constant $c > 0$ is small enough. We are interested in the curve $f(x,y) = 0$ which, in the coordinates (x, y_1), has the form $y_1 = \pm\sqrt{g(x^2) + c}$. From the properties: $g(0) = 0$; $g'(0) < 0$ for $0 < a < 2$; $g'(0) > 0$ for $a > 2$; $g'(0) = 0$ and $g''(0) < 0$ for $a = 2$, we can reproduce the graph of $g(z)$; it is shown in Fig. 2.15. From this we find the shape of the curve $f = 0$ on the plane of the variables x, y_1 (shown in Fig. 2.16). (On the plane of x, y variables this curve is some sense "kicked".) Let us note that the curve $f = 0$ has three components, one of which surrounds the point $(0, 0)$.

It turns out that the function

$$\Phi(x, y) = 1/f(x, y)$$

is the Dulac function for the van der Pol system in the domain $f > 0$. Indeed, we have

$$\operatorname{div}(v/f) = (f \cdot \operatorname{div} v - \dot{f})/f^2,$$

where

$$f \cdot \operatorname{div} v - \dot{f} = -a\left(\frac{2}{3}x^4 - cx^2 + c\right) =: M(x)$$

and $M(x) < 0$ for $0 < c < 8/3$ (Exercise 2.74).

We get two corollaries:

(a) $\dot{f}|_{f=0} > 0$, i.e., the vector field is directed into the interior of the domain $f > 0$;
(b) in the domain $f > 0$ the vector field Φv may have at most one limit cycle, moreover stable. Also v can have a unique limit cycle.

It remains to prove that there exists a limit cycle in $f > 0$. For this we construct the ring $\mathcal{R}$ satisfying the assumptions of Poincaré–Bendixson theorem. The inner boundary of this ring is a small closed component of the curve $f = 0$ around $(0,0)$. The outer boundary will be composed of pieces of the unbounded components of the curve $f = 0$ and of some arcs of the circle $x^2 + y^2 = R^2$ for the large radius R.

From property (a) it follows that, on the pieces of the boundary in $f = 0$, the field goes to the interior of $\mathcal{R}$. Next, from

$$\frac{\mathrm{d}}{\mathrm{d}t}\left(x^2 + y^2\right) = -2a\left(x^2 - 1\right)y^2,$$

it follows that $\dot{r} < 0$ is outside the band $|x| \leq 1$. But in this band we have

$$\frac{\mathrm{d}y}{\mathrm{d}x} = -a(x^2 - 1) - \frac{x}{y} = O(1) + O(1/R),$$

i.e., the increment of y (and hence the increment of r) is bounded by a constant independent of R. Now, it is easy to correct the corresponding pieces of the outer edge of the ring $\mathcal{R}$ so that the field also enters $\mathcal{R}$ (see Examples 2.23–2.24).

The above proof of the uniqueness of the limit cycle is due to L. Cherkas from Belarus.[g] In Exercise 4.31 in Chapter 4, we propose another proof of the uniqueness of the limit cycle in case the parameter $a > 0$ is small.

2.5. Drawing phase portraits on the plane

As already stated in Definition 2.1, the phase portrait of a vector field $v(x)$ on the plane is the foliation (or splitting) of the plane $\mathbb{R}^2$ into the phase curves of the field. The phase portrait elements are: singular points, closed phase curves, separatrices of singular points and behavior on infinity. We will briefly discuss them.

2.5.1. *Singular points*

They are divided into elementary and non-elementary. The elementary singular points can be subdivided into non-degenerate and degenerate ones.

Definition 2.40. An equilibrium point x_0 of the vector field $v(x)$, $x \in \mathbb{R}^n$, is called **non-degenerate** if $\det \frac{\partial v}{\partial x}(x_0) \neq 0$.

If the dimension $n = 2$, then x_0 is called **elementary** if at least one of the eigenvalues of the matrix $A = \frac{\partial v}{\partial x}(x_0)$ is non-zero. In this case, the point x_0 is:

- a **saddle**, if $\lambda_1 < 0 < \lambda_2$ for the eigenvalues $\lambda_{1,2}$ of the matrix A;
- a **stable node** (respectively, **unstable node**), if $\lambda_1, \lambda_2 < 0$ (respectively $\lambda_1, \lambda_2 > 0$);
- a **stable focus** (respectively, **unstable focus**), if $\lambda_{1,2} \in \mathbb{C} \setminus \mathbb{R}$ and $\operatorname{Re} \lambda_{1,2} < 0$ (respectively $\operatorname{Re} \lambda_{1,2} > 0$);
- a **saddle-node**, if $\lambda_1 = 0$, $\lambda_2 \neq 0$.

Local phase portraits near elementary singular points defined above are shown in Figs. 2.17 and 2.18.

---

[g]L. A. Cherkas, The Dulac function for polynomial autonomous systems on a plane, *Differential Equations* 33 (1997), 692–701 (translated from Russian).

Fig. 2.17.   Non-degenerate singular points.

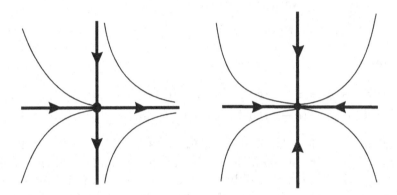

Fig. 2.18.   Saddle-nodes.

Remark 2.41. In the case of $\lambda_{1,2} \in i\mathbb{R} \setminus 0$, i.e., purely imaginary eigenvalues, the critical point x_0 may be a stable or unstable focus (more precisely, a **weak focus**) or a **center**; see Theorem 2.13.

Remark 2.42. The saddle-node concept given above is quite wide. The point is that we have the following model situations:

$$\dot{x} = x^{2k}, \qquad \dot{y} = \pm y, \qquad (2.31)$$

$$\dot{x} = \pm x^{2k+1}, \qquad \dot{y} = \pm y. \qquad (2.32)$$

In Section 3.3 in the following chapter, formulas (2.31)–(2.32) will be better justified. When $\dot{x} = x^s$ and $s > 2$, then the saddle-node is degenerate in some sense; we say that it has *codimension* $s - 1$.

From the topological perspective the local phase portrait does not depend on k. These portraits are shown in Fig. 2.18.

Remark 2.43 (Non-elementary singular points). Let us remind that here we have $\lambda_1 = \lambda_2 = 0$. Unfortunately, there is no satisfactory classification of such singularities. But there is an effective method of studying them. This is the method of **blowing-up of singularity**.

It consists of a simple operation of introducing the polar coordinates (r, φ) on the plane. So, for example, if we have

$$\dot{x} = ax^2 + bxy + cy^2 + \cdots, \quad \dot{y} = dx^2 + exy + fy^2 + \cdots, \quad (2.33)$$

then, in the polar variables, we get

$$\dot{r} = r^2(P(\varphi) + O(r)), \quad \dot{\varphi} = r(Q(\varphi) + O(r)), \quad (2.34)$$

where

$$P = a\cos^3\varphi + (b + d)\cos^2\varphi\sin\varphi + (c + e)\cos\varphi\sin\varphi + f\sin^3\varphi,$$
$$Q = d\cos^3\varphi + (e - a)\cos^2\varphi\sin\varphi + (f - b)\cos\varphi\sin\varphi - c\sin^3\varphi.$$

Note that the right-hand sides of Eq. (2.34) are zero at $r = 0$. Let us divide these right-hand sides by r; in the domain $r > 0$ the phase portrait will not change, only the "velocity" of the point along the phase curve will be different. It is the so-called *orbital equivalence* discussed in the following. (It may happen that $Q(\varphi) \equiv 0$, but then we divide by r^2.)

After this operation, we get a vector field on the cylinder $\{(r, \varphi)\} \simeq \mathbb{R} \times \mathbb{S}^1$ (from which important for us is the part $\{r \geq 0\}$) with isolated singular points on the circle $r = 0$. If the coefficients $a, \ldots, f$ are typical, then these singular points are already non-degenerate, i.e., elementary.

Otherwise, we repeat the procedure of blowing-up (combined with division) in the neighborhood of each non-elementary singular point $r = 0$, $\varphi = \varphi_0$ (with new coordinates $x = \varphi - \varphi_0$, $y = r$). It turns out that, if the original vector field is analytic, then, after the finite number of such blow-ups, we get a vector field on a surface M with elementary singular points; this procedure is called the **resolution of singularity**.[h]

---

[h]This is a difficult result, for details of which we refer to the monograph *The Monodromy Group* (Birkhäuser, 2006) by the first author.

Example 2.44. Consider vector field (1.1) from Example 1.2 (in Chapter 1), which we rewrite in the form

$$\dot{x} = y, \quad \dot{y} = -2x^3 - 4xy - ay\left(x^2 + y^2\right)^2 \tag{2.35}$$

(where $a = 1$) and denote it by v. We will show that its phase portrait is like in Fig. 1.2 in Chapter 1.

(a) We begin with the case $a = 0$, i.e.,

$$\dot{x} = y, \quad \dot{y} = -2x^3 - 4xy; \tag{2.36}$$

we denote it by v_0. Using the substitution $z = x^2$ one shows that its phase portrait in a neighborhood of $x = y = 0$ is qualitatively the same as in Fig. 1.2 in Chapter 1 (only it has a larger elliptic sector). Also, it turns out that the parabola $y = -x^2$ is invariant for field (2.36) and contains separatrices at the boundary of the hyperbolic sector (Exercise 2.78).

(b) Since the linearization matrix of field (2.35) at $(0,0)$ is nilpotent, this singularity is non-elementary. Let us apply the standard blowing-up, i.e., we put $x = r\cos\varphi$, $y = r\sin\varphi$. System (2.35) then becomes

$$\dot{r} = r\left\{\sin\varphi\cos\varphi - 4r\sin^2\varphi\cos\varphi - 2r^2\sin\varphi\cos^3\varphi - ar^4\sin^2\varphi\right\},$$

$$\dot{\varphi} = -\left\{\sin^2\varphi + 4r\sin\varphi\cos^2\varphi + 2r^2\cos^4\varphi + ar^4\sin\varphi\cos\varphi\right\}.$$

It has two singular points at the circle $r = 0$: $(r, \varphi) = (0,0)$ and $(0, \pi)$ (Fig. 2.19(a)).

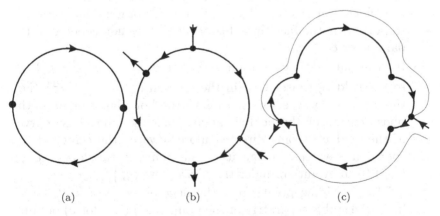

(a)   (b)   (c)

Fig. 2.19.   Resolution of singularity.

Near $(0,0)$ we introduce the local coordinates $x = r$, $y = \varphi$ (with abuse of notations). We find the local system

$$\dot{x} = xy + \cdots, \quad \dot{y} = -2x^2 - 4xy - y^2 + \cdots. \tag{2.37}$$

Note that the terms with the parameter a appear in the higher order terms. Again, the point $x = y = 0$ is non-elementary. Near $(0, \pi)$ we introduce the local coordinates $x = r$, $y = \varphi - \pi$ and get

$$\dot{x} = -xy + \cdots, \quad \dot{y} = -2x^2 + 4xy - y^2 + \cdots. \tag{2.38}$$

Let us apply the blowing-ups to the latter two systems. In case (2.37), we obtain the system

$$\dot{r} = -r^2 \sin \varphi \left(\sin^2 \varphi + 4 \sin \varphi \cos \varphi + \cos^2 \varphi \right) + O(r^3),$$
$$\dot{\varphi} = -2\sqrt{2} r \cos \varphi \sin^2 (\varphi + \pi/4) + O(r^2),$$

which should be divided by the common factor r. After that, we obtain a system with elementary singular points defined by $r = 0$ and $\varphi = \pm\frac{\pi}{2}, -\frac{\pi}{4}, \frac{3\pi}{4}$. The points $(0, \pm\pi/2)$ are saddles and the points $(0, -\pi/4)$ and $(0, 3\pi/4)$ are saddle-nodes, like at Fig. 2.19(b) (Exercise 2.78).

The blowing-up of system (2.38) is done analogously, and in the exceptional curve $r = 0$ we find a pair of saddled and a pair of saddle-nodes.

The complete resolution of the singularity of system (2.35) is presented at Fig. 2.19(c). The phase portrait outside the distinguished (closed and broken) curve is homeomorphic with the phase portrait of (2.35) outside the origin (and near it). Important is the fact that this portrait does not depend on the parameter a.

(c) It turns out that the addition of the vector field $-y \left(x^2 + y^2 \right)^2 \frac{\partial}{\partial y}$ to the field v_0 causes that in the domain $U = \left\{ y \geq -x^2 \right\}$ (i.e., the whole elliptic sector for v_0 with the boundary) we cross the phase curves of the field v_0 at an angle; so, the phase curves of the field v become directed more towards the center of the coordinate system. In conclusion, all points in the domain U go to $(0,0)$ in evolution under the phase flow $\left\{ g_v^t \right\}_{t>0}$.

The remaining points from the neighborhood of $(0,0)$ either lie in the stable separatrix of the point $(0,0)$ (i.e., for v) or enter the domain U after finite time under the flow g_v^t.

2.5.2. Closed phase curves

They correspond to the periodic solutions and were discussed in detail in the previous section.

2.5.3. Separatrices of singular points

We have the following definition.

Definition 2.45. A **separatrix of a singular point** x_0 of a vector field $v(x)$ is a phase curve of that field which "tends" to x_0 under a certain limit direction and is "distinguished" among such curves.

If there are no phase curves tending to x_0 under limit direction, then the singularity x_0 is called **monodromic**.

For example, for the saddles, the separatrices are components of the punctured stable submanifold $W^s \setminus x_0$ and of unstable $W^u \setminus x_0$; in total we have four separatrices.

In general case, when we divide a neighborhood of the origin into hyperbolic, elliptic, and parabolic sectors, separatrices lie on the boundaries of hyperbolic sectors.

In the case of monodromic singularity, one can define the Poincaré return map for a section $(S, x_0) \simeq (\mathbb{R}_+, 0)$ transversal to the phase curves.

In the case of a non-monodromic singular point, its neighborhood can be divided into sectors: **hyperbolic**, **parabolic** and **elliptic**. They are illustrated at Fig. 2.20. There is a theory associated with

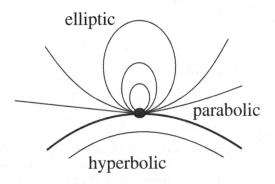

Fig. 2.20. Sectors.

them, which we will not deal with. To interested readers, we recommend the book of P. Hartman [5] and L. Perko [8].

Remark 2.46. An important element of the phase portrait of the vector field is the "fate" of the second end of a separatrix L.

It can land at another singular point x_1, usually in its parabolic sector. But it can also tend to a focus. Special is the case when L is a separatrix for both x_0 and for x_1; then we are dealing with the so-called **separatrix connection**.

The other end of a separatrix can also land on a limit cycle.

Particular and thoroughly investigated in Dynamic Systems Theory is the case when the other end of the separatrix L lands at the same point x_0. We have then the so-called **separatrix loop**. More on this topic can be found in Section 3.7.2 (in the next chapter) and in [9,10].

2.5.4. *Behavior at infinity*

There are multiple plane compactifications of the plane $\mathbb{R}^2$. One of them is the so-called **Poincaré compactification** (or the *Poincaré plane*). It relies on completing the plane with a circle by adding all "directions of infinity". Poincaré plane is diffeomorphic to the disk: $\mathbb{R}^2 \cup \mathbb{S}^1 \simeq D^2$ (see Fig. 2.21).[i]

Poincaré compactification is useful for studying the behavior of the phase curves of polynomial vector fields, i.e., the systems $\dot{x} = P(x,y)$, $\dot{y} = Q(x,y)$, whose right-hand sides P and Q are polynomials.

Then, in a neighborhood of the circle in infinity, one can introduce the polar type coordinates

$$x = \frac{1}{z}\cos\varphi, \quad y = \frac{1}{z}\sin\varphi.$$

We get a system of the form

$$\dot{z} = \frac{1}{z^k}\left(A(\varphi) + O(z)\right), \quad \dot{\varphi} = \frac{1}{z^l}\left(B(\varphi) + O(z)\right),$$

---

[i]In mathematics more widespread is the using the projective plane $\mathbb{RP}^2$. The difference between Poincaré compactification is that, in the projective plane, two antipodal directions in infinity are equivalent. Unfortunately, the projective plane is a non-oriented variety and cannot be drawn (in contrast to the Poincaré plane).

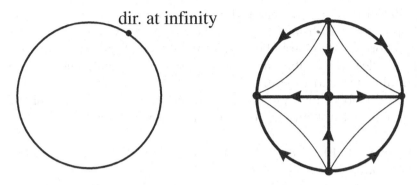

Fig. 2.21. Poincaré plane.

i.e., with the pole in the set $\{z = 0\}$ (circle in infinity). Multiplying the right side by $z^{\max(k,l)}$, which leads to an orbital equivalence in the domain $z > 0$, we get a regular vector field in a neighborhood of the infinity circle.

Let us note that the singular points $z = 0$, $\varphi = \varphi_0$ of the new field correspond to the situation where some phase curve of the system $\dot{x} = P$, $\dot{y} = Q$ tends to infinity (as $t \to +\infty$ or as $t \to -\infty$) under the limit direction φ_0.

Example 2.47. The linear vector field $\dot{x} = x$, $\dot{y} = -y$ leads to a phase portrait in the Poincaré plane as shown in Fig. 2.21.

2.5.5. *Orbital equivalence*

We begin with the corresponding definitions.

Definition 2.48. Two vector fields $v(x)$ on M and $w(y)$ on N are **orbitally equivalent** if they have the same phase portraits from a topological point of view. This means that there is a homeomorphism $h : M \longmapsto N$ such that h(phase curve of v) = (phase curve of w).

A vector field $v(x)$ on M is **orbitally structurally stable**, if every field $w(x)$ on M sufficiently close to v (in the appropriate topology) is orbitally equivalent to v.

It is easy to see that the above definition of orbital equivalence and orbital structural stability is weaker than the definition of equivalence and structural stability given in Definition 1.25 (in Chapter 1).

Recall that the Grobman–Hartman theorem (Theorem 1.24) and Proposition 1.26 from Chapter 1 show that two vector fields in the neighborhood of hyperbolic singular points with the same dimensions of a stable and unstable variety are orbitally equivalent. They are also orbitally structurally stable.

In the following, we present the necessary conditions for the orbital structural stability of a vector field on the plane.

Theorem 2.49. *If a field $v(x)$ in a domain $U \subset \mathbb{R}^2$ is orbitally structurally stable, then:*

- *its singular points are hyperbolic,*
- *its closed phase curves are hyperbolic limit cycles,*
- *there are no separatrix connections.*

Moreover, if the domain U is compact (e.g., Poincaré plane), then the above conditions for the orbital structural stability are also sufficient.

Proof. At first glance this theorem looks obvious. Indeed, the property of being non-hyperbolic for a singular point is "rare"; an additional condition is imposed. The same is true when any of the other conditions fails.

But we have to relate these analytic properties with some topological properties of the corresponding phase portraits. More precisely, to prove the necessity part of the theorem we should perturb a vector field v_0, for which one of the conditions fails, to a vector field v_ε with qualitatively different phase portrait. It is also clear that one can consider only local perturbations.

In the case of a separatrix connection, one can add to v_0 a small vector field in a direction transverse to the connection curve, so that the connection will become destroyed.

A similar push can be applied in the case of a non-hyperbolic periodic trajectory γ_0 of v_0. As a result the number of periodic trajectories will change. Here one works essentially with a perturbation of the Poincaré return map with fixed Poincaré section.

But the case of non-hyperbolic singular point is less obvious. There is no problem with perturbation of an elementary non-hyperbolic singular point, i.e., a saddle-node of codimension ≥ 1.

But in the case of non-elementary singular point p_0 of v_0, the perturbation should be adapted to the resolution process of this singularity; this is quite involved and we skip this part of the proof.

On the other hand, the sufficiency part of the proof is rather obvious, because in the compact phase space there are only finitely hyperbolic singular points with finitely many separatrices, and hyperbolic limit cycles.

Finally, we note that more precise approach to perturbations of non-generic vector fields is developed in the next chapter. □

2.6. Exercises

Exercise 2.50. Calculate the solution of the system $\dot{x} = y$, $\dot{y} = -\sin x$ (mathematical pendulum) with the initial condition $x(0) = 0$, $y(0) = 2$ (corresponding to $H = 1$).

Exercise 2.51. Similarly to the mathematical pendulum, find the first integral and sketch the phase curves for the *Duffing system*

$$\dot{x} = y, \quad \dot{y} = -x + x^3.$$

Calculate the solution with the initial condition $x(0) = 0$, $y(0) = 1/\sqrt{2}$.

Exercise 2.52. Calculate $\dot{r}$ and in $\dot{\varphi}$ in Example 2.4.

Exercise 2.53. Show that the eigenvalues of the matrix A in the transformation of Poincaré return $f(z) = Az + \cdots$ do not depend on the choice of section S to the periodic orbit γ.

Exercise 2.54. Prove formula (2.2). Show that $A(r, \varphi) = \sum_{j \geq 0} a_j (\varphi) r^j$ and $B(r, \varphi) = \sum_{j \geq 0} b_j (\varphi) r^j$, where $a_j(\varphi)$ and $b_j(\varphi))$ are homogeneous trigonometric polynomials of degree $j + 3$.

Exercise 2.55. Calculate the coefficient a_1 in Eq. (2.4) and prove that, if $a_1 = 1$, then $a_2 = 0$.
Hint: $\frac{dr}{d\varphi} = r^2 a(\varphi) + O(r^3)$, where the trigonometric polynomial $a(\varphi)$ is odd, i.e., $a(\varphi + \pi) = a(\varphi)$.

Exercise 2.56. Let $g(x) = b_1 x + b_2 x^2 + \cdots$ be an analytic diffeomorphism such that $b_1 < 0$ and let $f(x) = g \circ g(x) = a_1 x + a_2 x^2 + \cdots$. Prove that, if $\ln a_1 = 0$ then $a_2 = 0$ and, more generally, if $\ln a_1 = a_2 = \cdots = a_{2k-1} = 0$ then $a_{2k} = 0$.

Exercise 2.57. Assuming that on the right-hand side of Eq. (2.5) there are only quadratic terms (i.e., $D = E = \cdots = 0$) and that $c_3 = 0$, check the calculations leading to Eq. (2.8) for c_5.

Exercise 2.58. Prove that the divergence of a plane vector field written using complex variables as $\dot{z} = P(z, \bar{z})$ equals $2 \operatorname{Re} \frac{\partial P}{\partial z}$. Prove also that, if the divergence vanishes, then we have $\dot{z} = -2\mathrm{i} \frac{\partial H}{\partial \bar{z}}$, where $H(z, \bar{z})$ is the Hamilton function.

Exercise 2.59. Realize the change leading to real A, B, C in the case Q_2^R from Eq. (2.10).

Exercise 2.60. Prove that the function (2.11) is a first integral.

Exercise 2.61. Prove Eqs. (2.13)–(2.14).

Exercise 2.62. Check the computation of the order of tangency of the curves (2.20).

Exercise 2.63. Show that for a linear vector field the number of cycles is 0.

Exercise 2.64. Show that the ω-limit set $\omega(x)$ for the flow $\{g^t\}$ is closed and invariant, i.e., $g^t(\omega(x)) = \omega(x)$.

Exercise 2.65. Show that, in the proof of Proposition 2.22, the consecutive points x_j of intersection of the trajectory $\Gamma(x)$ of the point x with the section S form a monotone sequence, like in Fig. 2.7.

Exercise 2.66.

(a) Show that vector field (2.23) from Example 2.23 has the property that the square of the radius $r^2 = x^2 + y^2$ grows along the solution for small r.
(b) Prove inequalities (2.26)–(2.27).

Exercise 2.67. Prove that $\dot{L} < 0$ in Example 2.24.

Exercise 2.68. Calculate the index in $(0, 0)$ for the following fields:

(a) $\dot{x} = x^3 - 3xy^2$, $\dot{y} = y^3 - 3x^2y$;
(b) $\dot{x} = y + x^2$, $\dot{y} = x^3$;
(c) $\dot{x} = x^3 - 2y^3$, $\dot{y} = x(2x^2 - y^2)$.

Exercise 2.69. Show that $i_0 v = \text{sign} \det A$ for a germ $\dot{x} = Ax + \cdots$ in $(\mathbb{R}^n, 0)$ such that $\det A \neq 0$.

Exercise 2.70. Prove the formula $i_{x_0} v = (e - h)/2 + 1$, where e and h denote respectively the number of elliptic and hyperbolic sectors at the singular point x_0.

Exercise 2.71. Show that the vector field $\dot{x} = 1 - xy$, $\dot{y} = x$ does not have limit cycles.

Exercise 2.72. Show that the characteristic exponent of the periodic orbit defined in Definition 2.32 is well defined.

Exercise 2.73. Complete the analysis of the Jouanolou system in the case of odd s.

Exercise 2.74. Calculate $\text{div}(v/f)$ in the analysis of the van der Pol system.

Exercise 2.75. Sketch the phase portrait (in $\mathbb{R}^2$) for the system $\dot{x} = 1 + x^2 - y^2$, $\dot{y} = x + x^2 - 2xy$.

Exercise 2.76. Sketch the phase portrait for the system $\dot{x} = y^2 - 4x^2$, $\dot{y} = 4y - 8$.

Exercise 2.77. Sketch the phase portrait for the equations $\dot{z} = z^2$ and $\dot{z} = \bar{z}^2$, where $z = x + iy \in \mathbb{C} \simeq \mathbb{R}^2$.

Exercise 2.78.

(a) Prove the statements about system (2.35) in Example 2.44.
(b) Prove the statements about the blowing-ups of the vector fields (2.37) and (2.38).

Exercise 2.79. Resolve the following Bogdanov–Takens singularity:

$$\dot{x} = y + ax^2 + bxy + cy^2 + \cdots, \qquad \dot{y} = dx^2 + exy + fy^2 + \cdots,$$

with $d \neq 0$.

Chapter 3

Bifurcation Theory

3.1. Versality

According to Theorem 2.49 from Chapter 2 typical vector fields on a compact two-dimensional manifold M are orbitally structurally stable. If we denote by $\mathcal{X}$ the infinitely dimensional space of all vector fields on M (of given class and with a suitable topology about which we will not talk) then the subset Σ of the space $\mathcal{X}$, called the *bifurcation set*, consisting of fields which are not orbitally structurally stable should have codimension ≥ 1. It is be expected that this subset Σ is generally smooth, but it may have singular points (as in Fig. 3.1). The latter points should correspond to vector fields which have more complex singularities than those corresponding to typical points from Σ.

If we randomly choose a field from $\mathcal{X}$ then, with probability 1, it will lie outside of the set Σ. But a whole family $\{v_\lambda\}_{\lambda \in \mathbb{R}}$ of the vector field is a curve in $\mathcal{X}$ and can already cross the hypersurface Σ. We also expect that, for a typical family, the corresponding curve will cross the Σ hypersurface at a non-zero angle and at typical points of this hypersurface (see Fig. 3.1).

The bifurcation theory is concerned with the study of both the geometry of the bifurcation set Σ and the behavior of the multi-parameter families of the vector fields.

One should pay attention to one more aspect of this situation. On the space $\mathcal{X}$ acts the group $\mathcal{G}$ of orbital equivalences and the subset Σ is invariant with respect to that action. It is therefore necessary to link the analysis of 1-parameter families $\{v_\lambda\}$ to the action of

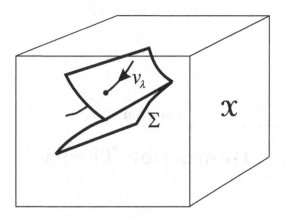

Fig. 3.1.   Bifurcation set.

group $\mathcal{G}$. V. Arnold in [6] introduced an order in this area and the following definitions are essentially due to him.[a]

Definition 3.1. Two families $\{v_\lambda\}_{\lambda \in J}$ and $\{w_\lambda\}_{\lambda \in J}$, $J \subset \mathbb{R}^k$, of the vector fields on M are **orbitally equivalent** if, for every $\lambda \in J$, the fields $v_\lambda(x)$ and $w_\lambda(x)$ are orbitally equivalent by homeomorphisms $h_\lambda(x)$, which depend in a continuous way on the parameter λ.

We say that a family $\{w_\nu\}_{\nu \in K}$, $K \subset \mathbb{R}^l$, is **induced** from a family $\{v_\lambda\}_{\lambda \in J}$ if there exists a continuous mapping $\varphi : K \longmapsto J$ such that

$$w_\nu = v_{\varphi(\nu)}.$$

A family $\{v_\lambda\}_{\lambda \in (\mathbb{R}^k, 0)}$, with the given v_0, is **versal** if any other family $\{w_\nu\}_{\nu \in (\mathbb{R}^l, 0)}$ with $w_0 = v_0$ is orbitally equivalent to a family induced from the family $\{v_\lambda\}$.

Example 3.2. The family $\dot{x} = \nu^2 + x^3$ is induced from the family $\dot{x} = \lambda + x^3$ by means of the change of parameters $\lambda = \varphi(\nu) = \nu^2$.

---

[a]This philosophy also works in other general situations. For example, when $\mathcal{X}$ is a space of functions f on a manifold and $\mathcal{G}$ is the group of diffeomorphisms h of the manifold with the action $f \longmapsto f \circ h$. Likewise, $\mathcal{X}$ can be a space of the diffeomorphisms f of a manifold and $\mathcal{G}$ the group of diffeomorphisms of the manifold with the action defined by the conjugation: $f \longmapsto h \circ f \circ h^{-1}$.

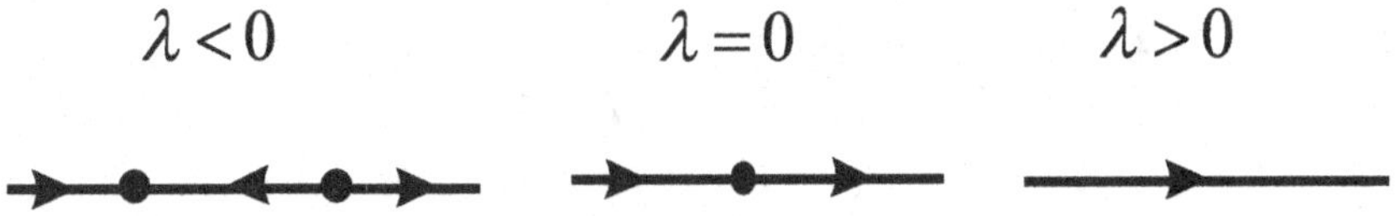

$$\lambda < 0 \qquad\qquad \lambda = 0 \qquad\qquad \lambda > 0$$

Fig. 3.2.   Phase portraits.

Example 3.3. Consider the model family

$$\dot{x} = v_\lambda(x) = \lambda + x^2. \tag{3.1}$$

The corresponding phase portraits are shown in Fig. 3.2.

Let us now take any family of the form

$$\dot{x} = w_\mu(x) = x^2 + \mu\tilde{w}(x,\mu) =: f(x,\mu), \tag{3.2}$$

where $\tilde{w}(x,\mu)$ is a smooth function in a neighborhood of $x = \mu = 0$. We claim that:

For every μ, the vector field w_μ has either 1 or 2 or 0 singular points in a neighborhood of $x = 0$.

To see this, consider the equation

$$g(x,\mu) = 0,$$

where $g(x,\mu) = f'_x(x,\mu)$. Since $f''_{xx}(0,0) \neq 0$, then $g'_x(0,0) \neq 0$, and by Implicit Function Theorem there exists a smooth function $\psi(\mu)$ such that the equation $g(\psi(\mu),\mu) \equiv 0$ is satisfied. This means that the point

$$x_\mu = \psi(\mu)$$

is the point of local minimum of the function $w_\mu = f(\cdot,\mu)$: $dw_\mu/dx(x_\mu) = 0$. We have three possibilities for the value of the field w_μ at point x_μ: (i) $w_\mu(x_\mu) > 0$, (ii) $w_\mu(x_\mu) < 0$, (iii) $w_\mu(x_\mu) = 0$. In case (i), the field w_μ has no equilibrium points, in case (ii) the field w_μ has two hyperbolic equilibrium points and in case (iii) there is one equilibrium point, a saddle-node (see Fig. 3.3).

So, for every μ, the phase portrait of the field w_μ is topologically equivalent to the phase portrait of the field v_λ for a corresponding λ. There is a natural question whether we can obtain $\lambda = \varphi(\mu)$ and homeomorphisms h_μ which realize an orbital equivalence of w_μ with $v_{\varphi(\mu)}$ and that the dependence of h_μ on μ is continuous. It turns out that yes; it means that family (3.1) is versal.

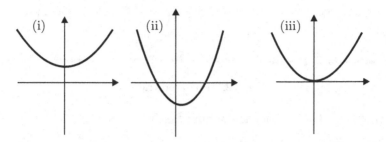

Fig. 3.3.   Right-hand sides.

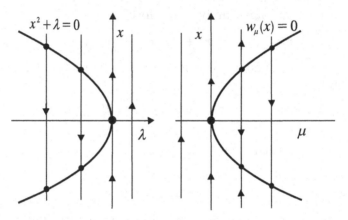

Fig. 3.4.   Bifurcations of model family and of typical family.

In order to convince ourselves, let us first consider the case where

(1) $\frac{\partial f}{\partial \mu}(0,0) \neq 0$ in Eq. (3.2). Then the curve of the equilibrium points

$$\Gamma : \{w_\mu(x) = f(x,\mu) = 0\}$$

of field w_μ, on the plane of the variables x, μ, is "vertical" and homeomorphic with a parabola (see Fig. 3.4). Also the curve of the equilibrium points $\Delta = \{\lambda = -x^2\}$ for the field v_λ is a parabola.

Depending on the sign of $f'_\mu(0,0)$ we put $\varphi(\mu) = \mu$ or $\varphi(\mu) = -\mu$; so, we can assume that both "parabolas" are oriented the same way. In this case, the constructions of h_μ's, i.e., of the homeomorphism of the plane

$$(\mu, x) \longmapsto (\mu, h_\mu(x)),$$

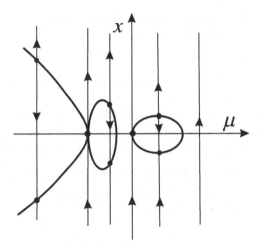

Fig. 3.5.   Bifurcation of non-generic family.

begins with the construction of the homeomorphism between the curves of the equilibrium points: $\Gamma \longmapsto \Delta$. Then we extend this homeomorphism continuously to the plane so that the vertical segments (at $\mu =$ const), outside the equilibrium points, go to the corresponding vertical segments. It is clear that it is possible to do so.

(2) In the degenerate case, when $f'_\mu(0,0) = 0$, the curve of the equilibrium points $\Gamma = \{f = 0\}$ can be very complex (see Fig. 3.5). But we know that the curve Γ has at most two points of intersection with each vertical line $\mu =$ const. Let us denote the intersection of Γ with such a straight line by Γ_μ. Then the change of parameters $\mu \longmapsto \varphi(\mu)$ is such that the parameters μ, for which $\#\Gamma_\mu = 2$, go to the left of $\lambda = 0$ and the parameters μ, for which $\#\Gamma_\mu = 0$, go to the right of $\lambda = 0$ (with continuity). Then, repeating arguments from case (1), we construct firstly a continuous transformation $(\mu, x) \longmapsto (\varphi(\mu), h_\mu(x))$ between the curves Γ and Δ and then extend it in a continuous way with the preservation of the verticality.

3.2.   Transversality

A mathematical tool for rigorous formulation of bifurcation theory and corresponding theorems is the transversality theory formulated by French mathematician R. Thom.

Let M be an n-dimensional manifold and let $B \subset M$ be its k-dimensional submanifold. Additionally, let A be an m-dimensional manifold and

$$f : A \longmapsto M$$

be a differential mapping (of sufficient class of differentiability). In the case of compact manifolds A and M in the space $C^k(A, M)$, a natural topology is introduced (which will not be specified).

Definition 3.4. We say that **the mapping f is transversal to the submanifold B** if, for every point $x \in A$ such that $y = f(x) \in B$, the following property holds:

$$f_* T_x A + T_y B = T_y M.$$

When $A \subset M$ is a submanifold and $f = id|_A$ is an embedding, then we say that A **is transversal to** B if the property

$$T_y A + T_y B = T_y M$$

holds for every point $y \in A \cap B$. We have the standard notations

$$f \pitchfork B \quad \text{and} \quad A \pitchfork B$$

for the transversality properties.

Example 3.5.

(1) Let $M = \mathbb{R}^2$ and $A \subset M$ and $B \subset M$ be smooth curves. Then $A \pitchfork B$ when curve A crosses curve B under a non-zero angle (see Fig. 3.6).

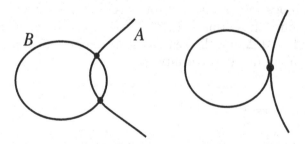

Fig. 3.6.   Curves in plane.

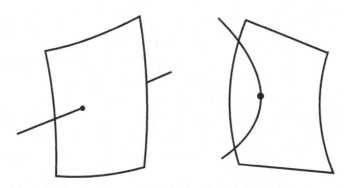

Fig. 3.7.   Curve and surface in space.

(2) Let $M = \mathbb{R}^3$, $A \subset M$ be a curve and $B \subset M$ be a surface. Figure 3.7 shows cases of transversality and non-transversality.

(3) The case with $M = \mathbb{R}^3$ and two surfaces $A \subset M$, $B \subset M$ is shown in Fig. 3.8.

(4) When $M = \mathbb{R}^3$ and $A \subset M$, $B \subset M$ are curves, then $A \pitchfork B$ if and only if the curves are disjoint.

(5) Let $M = \mathbb{R}^2 = \{(x,y)\}$, $A = \mathbb{R}^1 = \{t\}$ and $B = \{x = 0\}$ and let $f : A \longmapsto M$ be given by the formula $f(t) = (t^3, 0)$. Of course $f(t) \in B$ only for $t = 0$. But then $f_*(0) = f'(0) = 0$. So, $f_* T_0 A + T_{(0,0)} B = T_{(0,0)} B \simeq B \neq T_{(0,0)} M$. This example justifies the notion of the transversality of a mapping to a submanifold.

The following fundamental theorem is due to R. Thom.[b]

Theorem 3.6 (Thom). *Let M, A and B be fixed compact manifolds as above. Then the set of maps $f : A \longmapsto M$ that are transversal to B is open and dense in $C^r(A, M)$.*

This means that, on one hand, if a mapping f_0 is transversal to B then any mapping f_ε close to it is also transversal to B, and,

---

[b]The Thom transversality theorem, as well as its generalization given below, constitutes an important ingredient of his Catastrophe Theory. This theory includes, in part, the theory of singularities of mappings and functions as well as the Bifurcation Theory of Dynamical Systems.

It is worth adding that, in the case of the generalization of the Thom theorem on the case of non-compact manifolds, a special topology (the so-called Whitney topology) is introduced in the space of C^k class maps.

Fig. 3.8.   Surfaces in space.

on the other hand, if f_0 is not transversal then there exist maps f_ε, arbitrarily close to it, which is already transversal to B.

Proof.   It is easy to see that we can limit ourselves to the local situation when $A \subset \mathbb{R}^m$ and $M \subset \mathbb{R}^n$ are open subsets and

$$B = \{y_1 = \cdots = y_{n-k} = 0\} \cap M$$

is (locally) a subspace of codimension $n - k$ (Exercise 3.33(c)).

Then $f(x) = (f_1(x), \ldots, f_n(x))$ and

$$T_y B = \{q \in \mathbb{R}^n : q_1 = \cdots = q_{n-k} = 0\} \subset T_y M = \mathbb{R}^n.$$

If $f(x_0) \in B$, i.e., $f_1(x_0) = \cdots = f_{n-k}(x_0) = 0$, then f is transversal at x_0 to B if and only if the vectors $f_* \partial/\partial x_1, \ldots, f_* \partial/\partial x_m$ together with the vectors $\partial/\partial y_{n-k+1}, \ldots, \partial/\partial y_n$ span $\mathbb{R}^n$. (Here $\partial/\partial x_j$ and $\partial/\partial y_k$ are the base vectors in $\mathbb{R}^m$ and $\mathbb{R}^n$ respectively and $f_* = Df(x)$). For this, it suffices that the projections of the vectors $f_* \partial/\partial x_j$ onto the quotient space $T_y M/T_y B \simeq \mathbb{R}^{n-k}$ expand that space. This means that the matrix

$$C = \begin{bmatrix} \partial f_1/\partial x_1 & \cdots & \partial f_1/\partial x_m \\ \vdots & \ddots & \vdots \\ \partial f_{n-k}/\partial x_1 & \cdots & \partial f_{n-k}/\partial x_m \end{bmatrix}$$

has rank rk $C = n - k$.

We have two possibilities:

(1) $m < n - k$, then the rank of C is smaller than $n - k$ (which is impossible);

(2) $m \geq n - k$, then the property of rk $C \geq n - k$ is an open condition in the space of matrices.

In case (1), the transversality means that there is no intersection of $f(A)$ with B and it is an open condition in the mapping space. Also in case (2) it is not difficult to show that the condition of transversality is open (Exercise 3.33(a)).

The density property of transversality in case (1) is easy; one has to push $f(A)$ away from B. Locally, it is done by adding a small vector to $(f_1, \ldots, f_{n-k})$.

To prove the same property in case (2) we need to introduce additional concepts. Consider the local mapping $g : A \longmapsto \mathbb{R}^{n-k}$,

$$g(x) = (f_1(x), \ldots, f_{n-k}(x)).$$

Recall (see Definition 3.7) that x_0 is a critical point for g if $\mathrm{rk}\frac{\partial g}{\partial x}(x_0) = \mathrm{rk}\, C < n - k$, and the value $g(x_0)$ is called the critical value for g. We will use the classical Sard theorem (Theorem 3.8), which ensures that the set of critical values for g is "rare". According to Exercise 3.33(b), $f \pitchfork B$ if and only if 0 is not a critical value for g. Let z be a non-critical value for g and close to zero. We will perturb the mapping $f = (g, h)$ in the following way:

$$f_\varepsilon(x) = (g(x) - z, h(x)).$$

It is easy to see that f_ε is transversal to B (Exercise 3.33(a)). $\quad\square$

Definition 3.7. Let $g : \mathbb{R}^m \longmapsto \mathbb{R}^l$ be a differentiable mapping. We say that x_0 is a **critical point** for g if $\mathrm{rk}\frac{\partial g}{\partial x}(x_0)$ is not maximal. The value $g(x_0)$ at the critical point is called the **critical value** for h.

Theorem 3.8 (Sard). *The set of critical values for a differential mapping $g : \mathbb{R}^m \longmapsto \mathbb{R}^l$ of sufficiently high class of smoothness has zero Lebesque measure.*

Proof. Consider firstly the case $m = l = 1$.

It is possible that the critical values of g form a dense subset of $\mathbb{R}$. Anyway, we can cover each critical point x_j by a segment I_j of width $|I_j| \leq \varepsilon$ for arbitrarily small ε. Because $g'(x_j) = 0$, the length of the image $g(I_j)$ will be of order $O(\varepsilon^2) = O(\varepsilon)|I_j|$ (see Fig. 3.9). Therefore,

$$\left| g\left(\bigcup I_j\right) \right| \leq O(\varepsilon) \cdot \left|\bigcup I_j\right| \leq O(\varepsilon),$$

which tends to zero as $\varepsilon \to 0$.

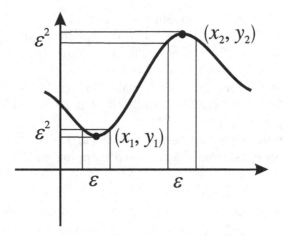

Fig. 3.9.   Sard theorem.

The same argument applies for any $m \leq l$ (Exercise 3.34).

Now, we pass to the proof in general situation. It uses induction with respect to m. Let $C = C(g)$ be the set of critical points of the map g. Assume also that the domain of definition of g is some bounded cube $U \subset \mathbb{R}^m$. The k-dimensional Lebesque measure of a Borel set $A \subset R^k$ is denoted by $|A|$.

Let $C_i \subset C$ be the set of points x at which vanish all partial derivatives of the components g_j of g order $\leq i$; thus $C \supset C_1 \supset C_2 \supset \cdots$. We shall prove that:

(i) $|g(C \backslash C_1)| = 0$,
(ii) $|g(C_i \backslash C_{i+1})| = 0$,
(iii) $|g(C_k)| = 0$ for some sufficiently large k.

(i) In the proof of the property (i), we can assume that $l \geq 2$ (for $l = 1$ we have $C = C_1$) and that $(\partial g_1 / \partial x_1)(x_0) \neq 0$ for $x_0 \in C$. Define the local diffeomorphism

$$(z_1, \ldots, z_n) = \Phi(x) = (g_1(x), x_2, \ldots, x_n),$$

from a neighborhood of x_0 to a neighborhood of $z_0 = \Phi(x_0)$, and the map $h = g \circ \Phi^{-1}$, which takes the following form:

$$h(z) = (z_1, h_2(z), \ldots, h_l(z)).$$

The latter map has the property $h_t = h|_{\{z_1 = t\}} : \{t\} \times \mathbb{R}^{m-1} \mapsto \{t\} \times \mathbb{R}^{l-1}$. The set $C(h)$ of critical points of the map h, i.e.,

$\Phi\left(C\left(g\right)\right)$, is defined by the sets of critical points of the maps h_t, $C\left(h\right) = \bigcup_t \left\{t\right\} \times C\left(h_t\right)$, and $h\left(C\left(h\right)\right) = \bigcup_t \left\{t\right\} \times h_t\left(C\left(h_t\right)\right)$. Now, we use the induction assumption and the Fubini theorem.[c] (Above the corresponding maps are local, like in other below maps.)

(ii) For proving property (ii), let $x_0 \in C_i \backslash C_{i+1}$. We can assume that $\left(\partial^{i+1} g_1 / \partial x_1 \partial x_{j_2} \ldots \partial x_{j_{i+1}}\right)\left(x_0\right) \neq 0$; of course, all derivatives of g_j of order $\leq i$ at p vanish. Again, define the local diffeomorphism

$$\left(z_1, \ldots, z_n\right) = \Phi\left(x\right) = \left(\partial^i g_1 / \partial x_{j_2} \ldots \partial x_{j_{i+1}}, x_2, \ldots, x_n\right),$$

and the map $h = g \circ \Phi^{-1}$. Denote also $h_0 = h|_{\{z_1=0\}} : \{0\} \times \mathbb{R}^{m-1} \longmapsto \mathbb{R}^l$. It turns out that:

$$\Phi\left(C_i\right) \subset \{z_1 = 0\} \quad \text{and} \quad \Phi\left(C_i\right) \subset C\left(h_0\right) \qquad (3.3)$$

(Exercise 3.35).

By induction assumption $\left|\Phi\left(g_0\right)\right| = 0$.

(iii) We claim that $\left|g\left(C_k\right)\right| = 0$ for $k \geq \frac{m}{l}$. For $x \in C_k$, we have

$$\left|g\left(x + h\right) - g\left(x\right)\right| < c\left|h\right|^{k+1}$$

for a constant $c > 0$ (depending on g and the cube U). We divide the cube U into subcubes with edges of small length $\varepsilon > 0$; there are about $\sim \varepsilon^{-n}$ such subcubes. If U_j is such a small cube containing $x \in C_k$ then the set $g\left(U_j\right)$ lies in a cube with the edge length $\leq c_1 \varepsilon^{k+1}$ (for some constant c_1); i.e., with the volume $\leq c_2 \varepsilon^{l(k+1)}$. It follows that

$$\left|g\left(C_k\right)\right| \leq c_3 \cdot \varepsilon^{-n} \cdot \varepsilon^{l(k+1)} \to 0$$

as $\varepsilon \to 0$.

From the proof it follows that the smoothness class of g should be $\geq \frac{m}{l} + 1$.

$\square$

---

[c]The Fubini theorem says that $\left|A\right| = \int \left|\{t\} \times A\right| dt$.

The following definition is needed to generalize Thom's theorem. Let

$$f : \mathbb{R}^m \longmapsto \mathbb{R}^n$$

be a mapping differentiable sufficiently many times. With it we can associate a series of geometric objects. The first is the graph

$$\{(x, f(x))\} \subset \mathbb{R}^m \times \mathbb{R}^n =: J^0(\mathbb{R}^m, \mathbb{R}^m).$$

Another is the graph of the derivative of the mapping $Df : x \longmapsto (y, p) = (f(x), \partial f / \partial x(x))$,

$$\{(x, f(x), \partial f / \partial x(x))\} \subset \mathbb{R}^m \times \mathbb{R}^n \times \mathbb{R}^{m \cdot n} =: J^1(\mathbb{R}^m, \mathbb{R}^n).$$

In general, the graph of the rth-order derivative mapping $x \longmapsto (f(x), Df(x), \ldots, D^r f(x))$ is a subset of a (rather large) space denoted by $J^r(\mathbb{R}^m, \mathbb{R}^n)$.

Definition 3.9. The spaces $J^r(\mathbb{R}^m, \mathbb{R}^n)$ are called the **spaces of jets of order** r from $\mathbb{R}^m$ to $\mathbb{R}^n$. With every mapping f of the above form is associated its rth-**jet**

$$x \longmapsto j^r f(x) = (x, f(x), Df(x), \ldots, D^r f(x)) \in J^r(\mathbb{R}^m, \mathbb{R}^n).$$

Analogously, if A and M are manifolds of dimension m and n, respectively, then one defines the **spaces** $J^r(A, M)$ **of the jets of order** r from A to M, and with each differential map $f : A \longmapsto M$ one associates its rth-**jet** $j^r f : A \longmapsto J^r(A, M)$.

Definition 3.10. If $C \subset J^r(A, M)$ is a submanifold, then we say that a mapping $f : A \longmapsto M$ is **transversal to** C (notation $f \pitchfork C$) when $j^r f \pitchfork C$.

Theorem 3.11 (Thom). *Let A, M and $C \subset J^r(A, M)$ be fixed. Then the set of mappings $f : A \longmapsto M$, which are transversal to C, is open and dense in the set of all such mappings.*

This proof in a large part repeats the proof of Theorem 3.6. The basic difference lies in the proof of density, more specifically, in the choice of the perturbation. Instead of the replacement $f(x) \to f(x) - z$ (where z is a "small" non-critical value for f), one takes the replacement $f(x) \to f(x) - z - \tilde{p}(x) - \tilde{q}(x, x) - \cdots - \tilde{s}(x, \ldots, x)$, where $\tilde{p}, \tilde{q}, \ldots, \tilde{s}$ are "small" linear map, homogeneous quadratic map, etc.

Example 3.12. Some natural conditions for a mapping can be interpreted as conditions for its transversality in jets. For example, the condition

$$\frac{\partial f}{\partial \lambda}(x, \lambda) \cdot \frac{\partial^2}{\partial x^2} f(x, \lambda) \neq 0 \quad \text{whenever} \quad f(x, \lambda) = \frac{\partial f}{\partial x}(x, \lambda) = 0$$

follows from the simultaneous transversality of the mapping

$$j^1 f : \mathbb{R}^2 = \{(x, \lambda)\} \longmapsto J^1(\mathbb{R}^2, \mathbb{R}) = \{(x, \lambda, y, p_x, p_\lambda)\}$$

to the two submanifolds

$$C_1 = \{y = 0\} \quad \text{and} \quad C_2 = \{y = p_x = 0\}.$$

Indeed, the transversality to C_1 means that $\frac{\partial y}{\partial x}(= \frac{\partial f}{\partial x}) \neq 0$ or $\frac{\partial y}{\partial \lambda}(= \frac{\partial f}{\partial \lambda}) \neq 0$ when $y(= f(x, \lambda)) = 0$. Whereas, the transversality to C_2 means that $\begin{vmatrix} f'_x & f'_\lambda \\ f''_{xx} & f''_{x\lambda} \end{vmatrix} \neq 0$ when $y(= f) = 0$ and $p_x(= f'_x) = 0$.

In problems of the Bifurcation Theory always, when conditions of a similar character like in Example 3.12 appear, we can assume that either they are satisfied with probability 1 or with the same probability cannot be satisfied. The second case occurs when $\dim A + \dim C < \dim J^r(A, M)$. This is the practical conclusion from the Thom's theorem.

3.3. Reductions

Assume that we have a vector field

$$\dot{x} = v(x) = Ax + \cdots$$

with the singular point $x = 0$. We can assume that the matrix A is block diagonal,

$$A = A^s \oplus A^u \oplus A^c, \tag{3.4}$$

where A^s (respectively, A^u and A^c) has eigenvalues with $\operatorname{Re} \lambda_j < 0$ (respectively, with $\operatorname{Re} \lambda_j > 0$ and with $\operatorname{Re} \lambda_j = 0$). The following result is a generalization of the Grobman–Hartman theorem.

Theorem 3.13 (Shoshitaishvili). *In the above situation, there exists a local homeomorphism* $h : (\mathbb{R}^n, 0) \longmapsto (\mathbb{R}^n, 0)$ *transforming the phase portrait of field* $v(x)$ *to the phase portrait of the following field:*

$$\dot{y}_1 = -y_1, \quad \dot{y}_2 = y_2, \quad \dot{y}_3 = w(y_3), \qquad (3.5)$$

where

$$w(y_3) = A^c y_3 + \cdots.$$

The proof of this theorem is technical and usually is skipped in the literature (see [11]). Therefore, we will not present it here. We only will draw from it some very practical applications.

Definition 3.14. The submanifold

$$W^c = \{y_1 = 0, \ y_2 = 0\}$$

(in terms of system (3.5)) is called the **center manifold**.

Shoshitaishvili theorem says that in the case of non-hyperbolic singular point the "interesting part" of dynamics takes place on the central manifold.

For a family of vector fields $v(x, \mu)$, $(x, \mu) \in (\mathbb{R}^n \times \mathbb{R}^k, (0, 0))$, we can treat μ as an additional variable, i.e., we have the system

$$\dot{x} = v(x, \mu), \quad \dot{\mu} = 0,$$

to which we apply Theorem 3.13 (with $h(x, \mu) = (h_0(x, \mu), \mu)$. We have the following **reductions to the center manifold**.

Proposition 3.15. *For the family* $\dot{x} = v(x, \mu)$ *with* $v(x, 0) = Ax + \cdots$, $A = A^s \oplus A^u \oplus A^c$, *there is a local homeomorphism* $(h_0(x, \mu), \mu)$ *realizing a topological equivalence of this family with the following family:*

$$\dot{z}_1 = -z_1, \quad \dot{z}_2 = z_2, \quad \dot{z}_3 = w(z_3, \mu).$$

The very existence of the center manifold is a generalization of the Hadamard–Perron theorem. It is possible to find it like in the proof of Theorem 1.19. In this case, the center manifold is not uniquely determined; it depends on the choice of the extension of the field (or of the diffeomorphism) to the whole $\mathbb{R}^n$.

But there is a way to describe W^c in a formal way. We look at it as the graph

$$W^c = \{x_1 = f(x_3), \quad x_2 = g(x_3) : x_3 \in E^c\}$$

(where the coordinates (x_1, x_2, x_3) are related to decomposition (3.4)), invariant for the field $v(x)$. On W^c, which is parametrized by $x_3 \in E^c$, we obtain the vector field

$$\dot{x}_3 = w(x_3).$$

Theorem 3.16 (Lyapunov–Schmidt reduction). *The mappings* $f : E^c \longmapsto E^s$ *and* $g : E^c \longmapsto E^u$ *have uniquely defined Taylor series at* $x_3 = 0$.

Proof. Since this reduction is essentially algebraic, we assume that the vector field $v(x)$ is complex analytic in $(\mathbb{C}^n, 0)$. Also, we firstly assume that the matrices A^s, A^u and A^c are diagonal; later we will remove this restriction (compare also the following proof of the Poincaré–Dulac theorem).

Let $\lambda_1, \ldots, \lambda_l \in \mathbb{C} \backslash i\mathbb{R}$ be the eigenvalues of the matrix $A^s \oplus A^u$ and $\nu_1 = i\omega_1, \ldots, \nu_m = i\omega_m \in i\mathbb{R}$ be the eigenvalues of A^c, and let $y_1, \ldots, y_l$ and $z_1, \ldots, z_m$ be the corresponding linear coordinates. Thus, we have the system

$$\dot{y}_j = \lambda_j y_j + \sum a_{j;\alpha,\beta} y^\alpha z^\beta, \quad j = 1, \ldots, l,$$

$$\dot{z}_k = \nu_k z_k + \sum b_{k;\alpha,\beta} y^\alpha z^\beta, \quad k = 1, \ldots, m,$$

where $y^\alpha = y_1^{\alpha_1} \cdots y_l^{\alpha_l}$, $z^\beta = z_1^{\beta_1} \cdots z_m^{\beta_m}$ for multi-indices $\alpha = (\alpha_1, \ldots, \alpha_l) \in (\mathbb{Z}_+)^l$ and $\beta = (\beta_1, \ldots, \beta_n)$ and the summations run over the pairs (α, β) such that $|\alpha| + |\beta| \geq 2$ (where $|\alpha| = \alpha_1 + \cdots + \alpha_l$).

We look for an invariant manifold W^c of the form

$$y_j = f_j(z), \quad f_j = \sum c_{j;\gamma} z^\gamma, \quad j = 1, \ldots, l, \tag{3.6}$$

where $|\gamma| \geq 2$. The invariance condition reads as follows:

$$\frac{\mathrm{d}}{\mathrm{d}t} (y_j - f_j(z)) |_{W^c} \equiv 0, \quad j = 1, \ldots, l. \tag{3.7}$$

The coefficients $c_{j;\gamma}$ are computed recurrently from the latter equations, by comparison of coefficients before z^γ. One can check that these recurrent equations are of the form (Exercise 3.36)

$$\{\lambda_j - (\nu, \gamma)\} c_{j;\gamma} + a_{j;0,\gamma} + \cdots = 0, \quad j = 1, \ldots, l, \quad |\gamma| = p, \quad (3.8)$$

where

$$(\nu, \gamma) = \nu_1 \gamma_1 + \cdots + \nu_m \gamma_m$$

and the dots denote terms expressed via the coefficients $a_{i;\alpha,\beta}$, $b_{r;\alpha,\beta}$ and $c_{s;\delta}$ with $|\delta| < |\gamma|$. Since $\operatorname{Re} \lambda_j \neq 0$ and $\operatorname{Re}(\nu, \gamma) = 0$, the above system of $l \times \#\{\gamma : |\gamma| = p\}$ equations has a unique solution.

In the case of existence of non-trivial Jordan cells for the matrix $A^s \oplus A^u$ and/or A^c, one can order suitably the variables y_j and/or z_k (and also the monomials z^γ) such that the corresponding linear system for $\{c_{j;\gamma} : j = 1, \ldots, l, |\gamma| = p\}$, analogous to (3.8), takes a triangular form with the diagonal entries $\lambda_j - (\nu, \gamma)$. $\square$

Remark 3.17.

(a) It turns out that generally the series of f and g are divergent (even if the vector field is analytic).

Consider the following example of Euler:

$$\dot{x} = x^2, \quad \dot{y} = y - x^2.$$

Here $W^u = \{x = 0\}$ is unstable manifold. We are looking for a center manifold $W^c = \{y = a_1 x + a_2 x^2 + a_3 x^3 + \cdots\}$. By putting such y into the system, we get $\left(a_1 + 2a_2 x + 3a_3 x^2 + \cdots\right) x^2 = \left(a_1 x + a_2 x^2 + a_3 x^3 + \cdots\right) - x^2$ (cf. Eq. (3.7)). This leads to the following recursion: $a_1 = 0$, $a_2 = 1$ and $a_{n+1} = n a_n$, with the solution $a_n = (n - 1)!$. Thus, the center manifold is given by the unique series

$$y = \sum_{n \geq 2} (n - 1)! x^n,$$

which is divergent.

(b) Moreover, as we have remarked above, the center manifold is not unique (in general). Consider for example, the saddle-node singularity $\dot{x} = x^2$, $\dot{y} = y$. Of course, the horizontal line is a natural center manifold. But also the set $\{y = 0, x \leq 0\} \cup \{y = Ce^{-1/x}, x > 0\}$, composed of two one-dimensional phase curves and of the singularity $(0, 0)$ can be regarded as W^c.

(c) Finally, as mentioned in Remark 1.17 (in Chapter 1), the reduction to the center manifold sometimes helps in the problem of stability of equilibrium points. As an example we revisit Exercise 1.38 (from Chapter 1) for $b = a < -1$, i.e., the system

$$\dot{x} = y + \sin x, \quad \dot{y} = a(x + y).$$

The center manifold we look for is $W^c = \{y = a_1 x + a_2 x^2 + \cdots\}$. The invariance condition

$$\frac{\mathrm{d}}{\mathrm{d}t}\left(y - a_1 x + a_2 x^2 + \cdots\right)|_{y=a_1 x + \cdots} \equiv 0$$

leads to the conditions for vanishing of the coefficients before initial powers of x: $(a - a_1)(1 + a_1) = 0$ (we choose $a_1 = -1$), $(a + 1) a_2 = 0$ (so $a_2 = 0$) and $(a + 1) a_3 - 1/6 = 0$; thus $W^c = \{y = -x + x^3/6(a + 1) + \cdots\}$. Since W^c is parametrized by x, we calculate

$$\dot{x}|_{W^c} = \left(\frac{1}{6(a+1)} - \frac{1}{6}\right) x^3 + \cdots = \frac{-a}{6(a+1)} x^3 + \cdots,$$

where $-a/6(a+1) < 0$. Since the other eigenvalue is negative, the point $(0, 0)$ is asymptotically stable.

Another tool useful in bifurcation theory, which also relies on (often divergent) formal series, is the next theorem. We consider germs of analytical vector fields

$$\dot{x} = Ax + O(|x|^2), \quad x \in (\mathbb{C}^n, 0), \tag{3.9}$$

such that, the matrix A is in (complex) Jordan form and has eigenvalues $\lambda_1, \ldots, \lambda_n$. Let us recall the standard basis $(e_1, \ldots, e_n)$ of $\mathbb{C}^n$ (therefore $x = \sum x_j e_j$) and notations $x^\alpha = x_1^{\alpha_1} \ldots x_n^{\alpha_n}$, $|\alpha| = \alpha_1 + \cdots + \alpha_n$.

Definition 3.18. We say that the eigenvalues satisfy the **resonance relation of type** $(j; \alpha)$, $j \in \{1, \ldots, n\}$, $\alpha = (\alpha_1, \ldots, \alpha_n) \in (\mathbb{Z}_+)^n$, if

$$\lambda_j = \alpha_1 \lambda_1 + \cdots + \alpha_n \lambda_n = (\alpha, \lambda).$$

Theorem 3.19 (Poincaré–Dulac). *Assume that we have a germ of analytic vector field (3.9). Then there exists a formal change*

$$x = y + O(|y|^2)$$

such that every component on the right is a formal power series in y, which leads to the system

$$\dot{y} = Ay + \sum c_{j;\alpha} y^\alpha e_j, \tag{3.10}$$

where the sum on the right-hand side of Eq. (3.10) runs over such pairs $(j; \alpha)$, $|\alpha| \geq 2$, that the resonance relation of type $(j; \alpha)$ holds.

Proof.　Assume firstly that the matrix A is diagonal. Reduction to the **Poincaré–Dulac normal form** (3.10) is realized by means of a series of changes of the form

$$x = y + \sum_{j,\alpha:|\alpha|=m} b_{j;\alpha} y^\alpha e_j, \tag{3.11}$$

i.e., we add to y_i homogeneous terms of the degree m. (We start with $m = 2$, then we take $m = 3$, etc.) It is easy to see that the inverse to (3.11) map has the form $y = x - \sum b_{j;\alpha} x^\alpha e_j + \cdots$ (Exercise 3.40).

　　Assume that only resonant terms are present in the part of field (3.9) of degree $\leq m - 1$ (inductive assumption). We want, by substituting (3.11), to eliminate the non-resonant terms of degree exactly m.

　　We have

$$\dot{x}_i = \dot{y}_i + \sum_{|\alpha|=m} b_{i;\alpha} \frac{\mathrm{d}}{\mathrm{d}t} y^\alpha. \tag{3.12}$$

We assume

$$\dot{y}_i = \lambda_i y_i + (\text{resonant terms of degree } \leq m) + h.o.t.$$

and hence

$$\sum b_{i;\alpha} \frac{\mathrm{d}}{\mathrm{d}t} y^\alpha = \sum b_{i;\alpha} \cdot (\lambda, \alpha) y^\alpha + h.o.t.$$

Now, on the left-hand side of Eq. (3.12), after substituting (3.11), we have

$$\lambda_i \left(y_i + \sum_{|\alpha|=m} b_{i;\alpha} y^\alpha \right) + \text{(res. terms of deg. } \leq m - 1)$$

$$+ \sum_{|\alpha|=m} a_{i;\alpha} y^\alpha + h.o.t.$$

On the right-hand side, we have

$$\lambda_i y_i + \text{(res. terms of deg. } \leq m - 1) + \sum_{|\alpha|=m} (\lambda, \alpha) b_{i;\alpha} y^\alpha + h.o.t.$$

So, comparing the homogeneous terms of degree m, we get the equations

$$\{(\lambda, \alpha) - \lambda_i\} \cdot b_{i;\alpha} = a_{i;\alpha}, \quad i = 1, \ldots, n, \quad |\alpha| = m. \tag{3.13}$$

It is clear from this that, if $(\lambda, \alpha) - \lambda_i \neq 0$ then one can compute $b_{i;\alpha}$, such that the corresponding non-resonant term $x^\alpha e_i$ becomes eliminated. Only resonant terms remain.

Equation (3.13) can be interpreted in terms of the following homological operator acting on formal vector fields $Z = \sum_{i;\alpha} b_{i;\alpha} x^\alpha \partial/\partial x_i$:

$$Z \mapsto \text{ad}_{V_0} Z = [V_0, Z], \tag{3.14}$$

where $V_0 = \sum_i \lambda_i \partial/\partial x_i$ is the leading (linear) part of the vector field (3.9) and $[\cdot, \cdot]$ denotes the commutator of two vector fields (Exercise 3.43). Of course, the latter operator is diagonal in the basis $\{x^\alpha \partial/\partial x_i\}$ with $(\lambda, \alpha) - \lambda_i$ as eigenvalues.

In the case of non-diagonal A, we can order the basis $x^\alpha \partial/\partial x_j$, $j = 1, \ldots, n$, $|\alpha| = m$, such that the above diagonal homological operator is replaced with a triangular operator with diagonal entries $(\lambda, \alpha) - \lambda_j$. □

Example 3.20. Let us consider the $1 : n$ *resonant node*

$$\dot{x}_1 = x_1 + \cdots, \quad \dot{x}_2 = nx_2 + \cdots,$$

i.e., with $\lambda_1 = 1$ and $\lambda_2 = n \in \mathbb{N} \backslash 1$. As we can see, the only resonant relation is $\lambda_2 = n\lambda_1 + 0 \cdot \lambda_2$. Thus, the Poincaré–Dulac normal form is

$$\dot{y}_1 = y_1, \quad \dot{y}_2 = ny_2 + cy_1^n$$

(Exercise 3.41).[d]

It turns out that in this case the change leading to the normal form is analytic (provided the germ was analytic).

To see this, we return to the proof of the Poincaré–Dulac theorem. Recall that we have there the operator $Z \mapsto \mathrm{ad}_{V_0} Z = [V_0, Z]$. In fact, it is a lowest degree part of the following **homological operator**:

$$Z \mapsto \mathrm{ad}_V Z = [V, Z],$$

which in order is a linear part, with respect to Z, of the application of the diffeomorphism $h = g_Z^1$ to the vector field V; we have

$$(\mathrm{Ad}_h)_* V = \left(Dh^{-1} \cdot V \right) \circ h = V + \mathrm{ad}_V Z + O(\|Z\|^2).$$

Here we equip vector fields $Z = \sum b_{j,\alpha} x^\alpha \partial/\partial x_j$ with the norms

$$\|Z\|_\rho = \sum_{j,\alpha} |b_{j,\alpha}| \, \rho^{|\alpha|};$$

so, if $\|Z\|_\rho < \infty$ then Z is analytic in a polydisc $D_\rho = \{|x_i| < \rho\} \subset \mathbb{C}^n$. Let $\mathcal{Z}_\rho$ be the Banach space of vector fields with finite this norm.

By normalizing the coordinates x_j we can assume that the constant c in the Poincaré–Dulac normal form is small. We can also assume that $V = V_0 + V_1 + V_1$, where $V_1 = cx_1^n \partial/x_2$ and the Taylor expansion of V_2 begins from high-order terms; V_2 is analytic in some polydisc D_ρ and $\|V_2\|_\rho$ is very small.

Then the operator $\mathrm{ad}_V = \mathrm{ad}_{V_0} + \mathrm{ad}_{V_1} + \mathrm{ad}_{V_2}$ is a small perturbation of the operator ad_{V_0} which is diagonal with the diagonal entries $(\alpha_1 - 1) + n\alpha_2$ and $\alpha_1 + n(\alpha_2 - 1)$. For sufficiently large $m = |\alpha| = \alpha_1 + \alpha_2$ these entries are bounded from below by Cm. Then the operator $(\mathrm{ad}_{V_0})^{-1}$, i.e., acting on vector fields with zero

---

[d]Into this case you can also include the situation with $n = 1$, that is, when the linear part is not diagonal. Also the following proof of the analyticity can be generalized to the higher dimensional case, when the convex hull (in $\mathbb{C}$) of the set of eigenvalues $\{\lambda_1, \ldots, \lambda_n\}$ is separated from $0 \in \mathbb{C}$; this case is called the Poincaré domain.

of sufficiently high order, is bounded. Also the operator $(\mathrm{ad}_V)^{-1}$ is bounded in $\mathcal{Z}_\rho$.

In the first step of the reduction, we apply $(\mathrm{Ad}_{h_1})_*$ with $h_1 = g_{Z_1}^1$ for $Z_1 = -(\mathrm{ad}_V)^{-1}V_2$. We get $(\mathrm{Ad}_{h_1})_* V = V_0 + V_1 + V_3$, where V_3 begins with high-order terms and $\|V_3\|_\rho = O(\|V_2\|_\rho^2)$. In the next step, we apply $(\mathrm{Ad}_{h_2})_*$ with $h_2 = g_{Z_2}^1$, $Z_2 = -(\mathrm{ad}_{V_0+V_1+V_3})^{-1}V_3$, etc. This is like in the Newtons method of recursive solution of a nonlinear equation.

In fact, in each step, we should slightly diminish the radius ρ of the polydisc of analyticity, but it can be done in a controlled way.

Example 3.21. For the *saddle-node*

$$\dot{x}_1 = \cdots, \quad \dot{x}_2 = x_2 + \cdots,$$

i.e., with $\lambda_1 = 0$ and $\lambda_2 = 1$, the resonant relations are of the form $\lambda_1 = k\lambda_1 + 0 \cdot \lambda_2$ and $\lambda_2 = k\lambda_1 + 1 \cdot \lambda_2$. Hence, the Poincaré–Dulac form as follows:

$$\dot{y}_1 = \sum_{k \geq 2} a_k y_1^k, \quad \dot{y}_2 = y_2 \left(1 + \sum_{k \geq 1} b_k y_1^k\right).$$

It turns out that this normal form is not analytic in general.

Example 3.22. For the $1 : -1$ *resonant saddle*

$$\dot{x}_1 = x_1 + \cdots, \quad \dot{x}_2 = -x_2 + \cdots,$$

i.e., with $\lambda_1 = 1$ and $\lambda_2 = -1$, the resonance relations are of the form $\lambda_1 = (k+1)\lambda_1 + k\lambda_2$ and $\lambda_2 = k\lambda_1 + (k+1)\lambda_2$. Hence, we have the following Poincaré–Dulac normal form:

$$\dot{y}_1 = y_1 \left\{1 + \sum a_k (y_1 y_2)^k\right\}, \quad \dot{y}_2 = -y_2 \left\{1 + \sum b_k (y_1 y_2)^k\right\}.$$

Also, this form is generally not analytic (Exercise 3.42).

Remark 3.23. There exist analogues of Theorems 3.13, 3.16 and 3.19 for germs of diffeomorphisms $f(x) = Ax + \cdots$ near a fixed point.

In Theorems 3.13 and 3.16, one assumes the splitting (3.4), where $|\lambda_j(A^s)| < 1$, $|\lambda_j(A^u)| > 1$ and $|\lambda_j(A^c)| = 1$.

In Theorem 3.19, the resonant relations between eigenvalues of the matrix A are of the following form (Exercise 3.47):

$$\lambda_j = \lambda_1^{k_1} \cdots \lambda_n^{k_n}.$$

3.4. Codimension one bifurcations

We will divide the bifurcations of autonomous vector fields into local and non-local ones.

Local bifurcations take place in a neighborhood of the singular point $x = 0$, i.e., singular when the parameter value is bifurcational. More precisely, we have a family

$$\dot{x} = v(x; \mu)$$

such that

$$v(x, 0) = Ax + \cdots.$$

In the case of typical 1-parameter families, the violation of the hyperbolic condition (i.e., $\operatorname{Re} \lambda_j(A) \neq 0$ for all the eigenvalues) occurs in two cases:

(1) $\lambda_1(A) = 0$ and $\operatorname{Re} \lambda_j(A) \neq 0$ for $j > 1$; this is a **saddle-node bifurcation**.

(2) $\lambda_1 = \bar{\lambda}_2 = i\omega \in i\mathbb{R} \setminus 0$ and $\operatorname{Re} \lambda_j \neq 0$ for $j > 2$; this is a **birth of the limit cycle bifurcation** or the **Andronov–Hopf bifurcation**.

We have three **non-local bifurcations** associated with a periodic orbit γ for a field $v(x, 0)$. Let λ_j be the eigenvalues of the linear part of the Poincaré return map. Recall that the hyperbolicity condition means that $|\lambda_j| \neq 1$ for all j. Thus, typical codimension 1 bifurcations may occur in the following cases:

(3) $\lambda_1 = 1$ and $|\lambda_j| \neq 1$ for $j > 1$; this is a **saddle-node bifurcation for periodic orbit**.

(4) $\lambda_1 = -1$ and $|\lambda_j| \neq 1$ for $j > 1$; this is a **period doubling bifurcation**.

(5) $\lambda_1 = \bar{\lambda}_2 \in \mathbb{S}^1 \setminus \{1, -1\}$. Here the cases where $\lambda_1 = e^{2\pi i k/m}$ is a root of unity of degree $m = 1$ are called **resonances**. Moreover, when

$m = 1, 2, 3, 4$, we are talking about **strong resonances**; in fact, the strong resonances should be treated rather as codimension two bifurcations (see [6]).

Next, there are two **non-local bifurcations** in $\mathbb{R}^2$ related with connections of separatrices (see Fig. 3.18):

(6) A **separatrix connection** between different saddles.
(7) A **separatrix loop** of one saddle.

Finally, in higher dimensional phase spaces more complicated non-local bifurcations associated with intersections of invariant manifolds of singular points and/or limit cycles are possible. They lead to new phenomena, like a chaotic dynamics (see Chapter 5). We do not study them here.

3.5. Saddle-node bifurcation

We have a 1-parameter family of vector fields

$$\dot{x} = v(x; \mu) = v_\mu(x), \quad (x, \mu) \in (\mathbb{R}^n \times \mathbb{R}, 0 \times 0).$$

We impose the following conditions:

(1) $v(0; 0) = 0$, therefore $v(x; 0) = Ax + \cdots$.
(2) The matrix A has one zero eigenvalue, $\lambda_1 = 0$ (in particular, $\det A = 0$) and $\operatorname{Re} \lambda_j \neq 0$ for $j > 1$. Thus, it can be assumed that A has the block form

$$A = \begin{bmatrix} 0 & 0 \\ 0 & B \end{bmatrix}.$$

(3) Let (x, y) be the linear coordinate system associated with the above A. We can rewrite our system as

$$\dot{x} = f(x, y; \mu), \quad \dot{y} = By + \cdots,$$

where $f(0, 0; 0) = 0$ and $f'_x(0, 0; 0) = 0$. The next assumption says that

$$f''_{xx}(0, 0; 0) \neq 0.$$

(4) The last assumption is that

$$f'_\mu(0, 0; 0) \neq 0$$

(Exercise 3.48).

Remark 3.24. The above conditions are conditions in the space $J^2 = J^2(\mathbb{R}^n \times \mathbb{R}, \mathbb{R}^n)$ of jets of order 2. These are conditions for transversality with respect to certain submanifolds in J^2. By Theorem 3.11, a typical $v(x; \mu)$ satisfies these conditions (see also Example 3.12).

Theorem 3.25. *If the above conditions are satisfied, then the family $\{v_\mu\}$ is versal. In particular, it is equivalent to one of the families of the form*

$$\dot{x} = \lambda \pm x^2, \quad \dot{y}_1 = -y_1, \quad \dot{y}_2 = y_2.$$

Proof. From the reduction to the center manifold theorem we can assume that we have a system of the form

$$\dot{x} = f(x, \mu), \quad \dot{y}_1 = -y_1, \quad \dot{y}_2 = y_2.$$

We have the following properties directly from conditions (1)–(4):

(1) $f(0, 0) = 0$,
(2) $f'_x(0, 0) = 0$,
(3) $f''_{xx}(0, 0) \neq 0$,
(4) $f'_\mu(0, 0) \neq 0$.

Further proof is exactly as in Example 3.3.

Figure 3.10 shows the saddle-node bifurcation for a family of two-dimensional vector fields. $\qquad\square$

Remark 3.26. Sometimes bifurcations such that an equilibrium point does not move, and does not disappear, are considered. They are called the **transcritical saddle-node bifurcations**. In one dimension, the model transcritical bifurcation is as follows:

$$\dot{x} = x(\nu \pm x). \qquad (3.15)$$

Of course, they are induced from the versal family from Theorem 3.25.

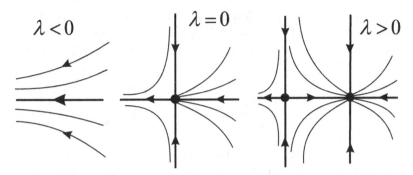

Fig. 3.10.   Saddle-node bifurcation in plane.

3.6.   Andronov–Hopf bifurcation

We have a 1-parameter family of vector fields

$$\dot{x} = v(x;\mu) = v_\mu(x), \quad (x,\mu) \in (\mathbb{R}^n \times \mathbb{R}, 0 \times 0).$$

We impose the following conditions:

(1) $v(0;0) = 0$, therefore $v(x;0) = Ax + \cdots$.

(2) $\lambda_1 = \bar{\lambda}_2 = i\omega \in i\mathbb{R} \setminus 0$ and $\operatorname{Re}\lambda_j \neq 0$ for $j > 2$. This implies $\det A \neq 0$ and by the Implicit Function Theorem the equation $v(x;\mu) = 0$ has the solution $x = x_0(\mu)$, which corresponds to the equilibrium point. Then we move this equilibrium point to the beginning of the coordinate system: $x \longmapsto x - x_0(\mu)$. Now, we have the system

$$\dot{x} = A(\mu)x + \cdots.$$

(3) The next assumption is that

$$\frac{\partial}{\partial \mu} \operatorname{Re} \lambda_{1,2}(\mu)|_{\mu=0} \neq 0.$$

(4) The last assumption is based on the Poincaré–Dulac normal form for $\mu = 0$. In the complex domain, we have $\lambda_1 = -\lambda_2$ and according to Example 3.22 the normal form becomes

$$\dot{z}_1 = i\omega z_1 \Big\{ 1 + \sum a_j (z_1 z_2)^j \Big\}, \quad \dot{z}_2 = -i\omega z_2 \Big\{ 1 + \sum b_j (z_1 z_2)^j \Big\},$$

where $z_{1,2} = y_1 \pm iy_2 = x_1 \pm ix_2 + \cdots$ are (formal) complex variables after restriction to the central manifold. Since the starting field is real, then $z_2 = \bar{z}_1$ and the above equations are conjugated one to another. In real variables $y_1 = x_1 + \cdots$ and $y_2 = x_2 + \cdots$, we get the following **Poincaré–Dulac normal form for a focus**:

$$\dot{y}_1 = y_1\left(c_3 r^2 + c_5 r^4 + \cdots\right) - \omega y_2\left(1 + d_3 r^2 + \cdots\right),$$
$$\dot{y}_2 = \omega y_1\left(1 + d_3 r^2 + \cdots\right) + y_2\left(c_3 r^2 + c_5 r^4 + \cdots\right),$$

where $r^2 = y_1^2 + y_2^2$. Here $c_3, c_5, \ldots$ are proportional to the Poincaré–Lyapunov focus quantities from Definition 2.12 (Exercise 3.49).

The last non-degeneracy condition says that

$$c_3 \neq 0.$$

The following statement is also called the **limit cycle creation theorem** (or the **Andronov–Hopf theorem**) and is perhaps the most known theorem of the bifurcation theory.

Theorem 3.27 (Andronov–Hopf). *If the above conditions are true then the family $\{v_\mu\}$ is versal. In particular, it is equivalent to the family*

$$\dot{x}_1 = x_1(\lambda \pm r^2) - \omega x_2, \quad \dot{x}_2 = \omega x_1 + x_2(\lambda \pm r^2), \quad \dot{y}_1 = -y_1, \quad \dot{y}_2 = y_2,$$

$r^2 = x_1^2 + x_2^2$, *or to the (equivalent) family*

$$\dot{z} = (\lambda + i\omega)z \pm z\,|z|^2, \quad \dot{y}_1 = -y_1, \quad \dot{y}_2 = y_2, \qquad (3.16)$$

$z = x_1 + ix_2 \in \mathbb{C} \simeq \mathbb{R}^2$.

Remark 3.28. Up to the change of orientation of the plane (e.g., $(x_1, x_2) \longmapsto (x_1, -x_2)$), we can assume that the frequency $\omega > 0$. With this assumption, we have two local bifurcations of families (3.16) corresponding to two values $c_3 = 1$ and $c_3 = -1$. They are shown in Figs. 3.11 and 3.12, respectively. The difference between these drawings is of great practical importance.

In Fig. 3.12, we observe the so-called **unsafe loss of stability** (or the so-called **subcritical bifurcation**). Indeed, for $\lambda < 0$ the equilibrium point is stable (although the "basin" of its attraction

$\lambda < 0$ $\lambda = 0$ $\lambda > 0$

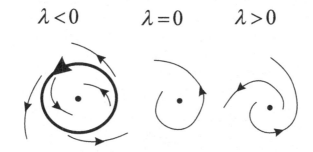

Fig. 3.11. Sharp loss of stability.

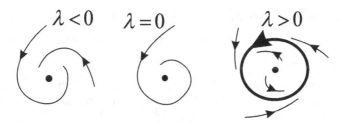

Fig. 3.12. Mild loss of stability.

shrinks) and for $\lambda \geq 0$ the equilibrium point becomes "globally" unstable (the system becomes completely destroyed).

Figure 3.13 shows the so-called **safe loss of stability** (or the **supercritical bifurcation**). For $\lambda \leq 0$ the equilibrium point is "globally" stable and for $\lambda > 0$ it loses stability. But, for $\lambda > 0$, there appears a stable limit cycle located near the equilibrium point. So, the system (e.g., mechanical) begins to oscillate slightly around the equilibrium position.

Proof of Theorem 3.25. Like in the proof of the saddle-node bifurcation theorem, we firstly reduce the problem to the two-dimensional situation (on the center manifold).

Modifying slightly the proof of the Poincaré–Dulac theorem, we bring the whole family to the following normal form, modulo terms of order ≥ 4:

$$\dot{z} = (c_1(\mu) + i\omega(\mu)) z + (c_3(\mu) + id_3(\mu)) |z|^2 z + O(|z|^4), \quad (3.17)$$

$z = y_1 + iy_2$, where $c_1(0) = 0$, $c_1'(0) \neq 0$ and $c_3(0) \neq 0$ (Exercise 3.50). We can easily assume that $c_1(\mu) = \mu$, and, passing to the

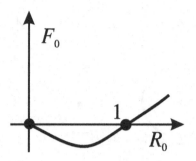

Fig. 3.13.   Function F_0.

polar coordinate system, $z = re^{i\varphi}$, we can write

$$\dot{r} = r\left(\mu + c_3 r^2 + O(r^3)\right), \quad \dot{\varphi} = \omega + O(r^2) \qquad (3.18)$$

(Exercise 3.51).

For system (3.17) we define the Poincaré return map $P : \{\varphi = 0\}$ $\longmapsto \{\varphi = 2\pi\}$ as in Fig. 2.5 in Chapter 2. By rescaling, we can put $c_3 = \pm 1$; for simplicity, we assume that $c_3 = 1$.

We divide the further analysis into three cases.

(1) For $\mu \geq 0$, we have $\dot{r} > 0$ (for $r > 0$), that is $P(r) > r$ and there are no limit cycles.

(2) Let $\mu < 0$. Consider the domain $\{0 \leq r \leq 2\sqrt{-\mu}\}$. Let us make the following normalization:

$$r = \tau R, \quad \mu = -\tau^2.$$

Then, in the domain $\{0 \leq R \leq 2\}$, we get the system

$$\dot{R} = \tau^2 R\left(-1 + R^2 + O(\tau)\right), \quad \dot{\varphi} = \omega + O(\tau^2) \qquad (3.19)$$

with the small parameter τ. Now, it is easy to calculate the Poincaré map. Note that the solution of system (3.19) with the initial condition $R(0) = R_0$, $\varphi(0) = 0$, satisfies $R(t) = R_0 + O(\tau)$. Therefore,

$$P(R_0) - R_0 = \int_0^{2\pi} \frac{\mathrm{d}R}{\mathrm{d}\varphi} d\varphi = \frac{\tau^2}{\omega} \int_0^{2\pi} \frac{R(-1 + R^2 + O(\tau))}{1 + O(\tau)} \mathrm{d}\varphi$$

$$= \tau^2 \frac{2\pi}{\omega} R_0(R_0^2 - 1 + O(\tau)).$$

Let us denote $F(R_0, \tau) = (P(R_0) - R_0)/\tau^2 = F_0(R_0) + O(\tau)$, where the graph of function $F_0(R_0) = \pi R_0(R_0^2 - 1)/\omega$ is shown in Fig. 3.13. We see that the equation $F(R_0, \tau) = 0$ has exactly two simple solutions $R_0 = 0$ and $R_0 = 1 + O(\tau)$ (Exercise 3.52). The first of them corresponds to the equilibrium point and the second to the unstable limit cycle close to the circle $\{r = \sqrt{-\mu}\}$.

(3) For $\mu < 0$, consider the domain $\{2\sqrt{-\mu} < r < \varepsilon\}$ for small $\varepsilon > 0$ (independent of μ). From Eq. (3.18), it is easy to see that here $\dot{r} > 0$ and there are no limit cycles.

Thus, we see that in each of the above three situations the phase portraits are "qualitatively" the same as for the model family (3.16). It would be desirable to construct the family $\{h_\mu\}$ of local homeomorphisms that implements the topological orbital equivalences of the respective phase portraits. It is a quite tedious task (if you take it seriously) and we omit it (even in [11] this is skipped). □

Remark 3.29. E. Hopf in his original work proved a more general result than Theorem 3.27.

Namely, he deleted the assumption that $c_3 \neq 0$, but kept the assumption $\partial \operatorname{Re} \lambda_{1,2} / \partial \mu \neq 0$ (see [12]). He showed the existence of a 1-parameter family γ_ν of periodic solutions for vector fields $v_{\psi(\nu)}(x)$, where $(\mathbb{R}_+, 0) \ni \nu \longrightarrow \mu = \psi(\nu)$ is a smooth mapping.

This statement is called the **Hopf bifurcation theorem**.

For example, for the family

$$\dot{z} = (\mu + \mathrm{i}) z = v_\mu(z, \bar{z}),$$

we have a family periodic solutions $\gamma_\nu = \{|z|^2 = \nu\}$ for one field $\dot{z} = \mathrm{i}z = v_0(z, \bar{z})$, i.e., $\psi(\nu) \equiv 0$.

To prove the Hopf theorem we can assume the system $\dot{z} = (\mu + \mathrm{i}\omega) z + O(|z|^3)$, $\omega = \omega_0 + O(|\mu|)$, and, in the polar coordinates, we get

$$\mathrm{d}r/\mathrm{d}\varphi = r \left(\mu/\omega_0 + O\left(r^2 + \mu^2\right)\right).$$

The Poincaré map $P(r_0, \mu)$ is the solution at $\varphi = 2\pi$ of this equation with the initial condition $r(0; r_0, \mu) = r_0$. We find

$$g(r_0, \mu) := \{P(r_0, \mu) - r_0\} / r_0 = (2\pi/\omega_0) \mu + O\left(r_0^2 + \mu^2\right)$$

with $\partial g/\partial \mu (0, 0) \neq 0$. Next, we apply the Implicit Function Theorem.

Arnold [11] often underlined that, in the case of $c_3 \neq 0$, the appropriate bifurcations were studied also by A. Andronov [10]. For this reason, the bifurcation from Theorem 3.27 is called the **Andronov–Hopf bifurcation.**

Example 3.30 (FitzHugh–Nagumo model). We revisit Example 2.24, i.e.,

$$\dot{x} = a + x - y - x^3/3, \quad \dot{y} = b + cx + dy.$$

Again, we replace the parameter b with the first coordinate of the position of one singular point (x_0, y_0) in the formula $b = ad + (d - c)x_0 - dx_0^3/3$ (and $y_0 = a + x_0 - x_0^3/3$). Let

$$u = x - x_0, \quad v = y - y_0.$$

We get the system

$$\dot{u} = \left(1 - x_0^2\right)u - v - x_0 u^2 - u^3/3, \quad \dot{v} = cu - dv.$$

The characteristic polynomial of the linearization matrix A_0 at $u = v = 0$ is $P(\lambda) = \lambda^2 + \left(x_0^2 + d - 1\right)\lambda + c - d + dx_0^2$. We assume that $c - d + dx_0^2 > 0$ and

$$\mu := \left(1 - d - x_0^2\right)/2 \approx 0;$$

so, μ will be a small parameter corresponding to the real part of the complex eigenvalues of A_0.

In order to demonstrate the existence of Andronov–Hopf bifurcation, we need to compute the third focus quantity when $\mu = 0$, i.e., $x_0^2 = 1 - d$. We use complex coordinates. Let the eigenvalues be $\lambda_{1,2} = \pm i\omega = \pm i\sqrt{c - d + dx_0^2} = \pm i\sqrt{c - d^2}$; thus

$$c = d^2 + \omega^2.$$

Then the complex variable

$$z = cu - dv + i\omega v$$

becomes the eigenfunction of A_0 with the eigenvalue $i\omega$. We have $u = \frac{1}{2c\omega}\{(\omega - id)z + (\omega + id)\bar{z}\}$ and we get the equation

$$\dot{z} = i\omega z - cx_0 u^2 - cu^3/3 = \omega\left\{iz + Az^2 + Bz\bar{z} + \cdots + G\bar{z}^3\right\}$$

(cf. Eq. (2.5) from the previous chapter) where

$$A = \frac{x_0}{4c\omega^3}\left(2\mathrm{i}d\omega + d^2 - \omega^2\right), \quad B = -\frac{x_0}{2\omega^3}, \quad E = -\frac{1}{8c\omega^4}\left(\omega - \mathrm{i}d\right)$$

and other coefficients are not important. Using Proposition 2.14, we get

$$c_3 = \operatorname{Re} E - \operatorname{Im} AB = -\frac{1}{8c\omega^3} + \frac{x_0^2 d}{4c\omega^5} = \frac{2d - 2d^2 - \omega^2}{4c\omega^5}.$$

It is non-zero for generic values of the parameters.

Example 3.31 (Zhukovskii plane model). Let a plane flies with speed v (which can vary) and is raised at an angle Θ with respect to the horizontal (see Fig. 3.14). The following forces act on the plane: the thrust force F_{thr} directed along the plane, the gravity force mg directed downwards and the air drag force proportional to v^2. We split the gravity force into components along the plane (causing loss of speed) and in the perpendicular direction (causing downward rotation). So, we have the following pair of equations: $m\dot{v} = F_{\mathrm{thr}} - mg\sin\theta - c_1 v^2$ and $mv\dot{\theta} = -mg\cos\theta + c_2 v^2$. Here the constants c_1 and c_2 depend on several factors, which we will not

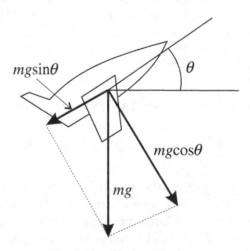

Fig. 3.14. Zhukovskii plane.

specify (see [13]). After suitable normalization $y = \text{const} \cdot v$, we get the following 2-parameter family of vector fields:

$$\dot{\theta} = (y^2 - \cos\theta)/y, \quad \dot{y} = \lambda - \mu y^2 - \sin\theta, \qquad (3.20)$$

$-\pi/2 \le \theta \le \pi/2$, $y \ge 0$, with the pole along $\{y = 0\}$.

Let us firstly consider the case $\lambda = 0$, which corresponds to the glider model. After multiplying by y (rescaling the time), we get the regular field

$$V = (y^2 - \cos\theta)\partial/\partial\theta - y(\sin\theta + \mu y^2)\partial/\partial y. \qquad (3.21)$$

For $\mu = 0$, we have a Hamiltonian system with the first integral

$$F = y^3/3 - y\cos\theta \qquad (3.22)$$

and the equilibrium points $S_{\pm} = (\pm\pi/2, 0)$ and $C = (0, 1)$. For $\mu > 0$, we get $\text{div}V = -2\mu y < 0$, meaning the function $\Phi = y$ is a Dulac function for the field (3.20). It is easy to see that, for $\mu = 0$, the glider's movement is periodic (oscillating around the C center) and for $\mu > 0$ the corresponding critical point C (bifurcating from $(0, 1)$) is a globally stable focus (Exercise 3.53). This means that the glider's movement is stabilizing.

For the general family (3.21) with small λ and μ, the possible limit cycles can be investigated using the so-called Abelian integrals method (see Section 4.1.3 in the next chapter). The increment ΔF of the first integral along the trajectory of system (3.21) (from section to section) is approximately

$$\int \dot{F}dt = \int (y^2 - \cos\theta)(\lambda - \mu y^2)\,dt \approx I(c),$$

$$I(c) = \int_{F=c} (\lambda - \mu y^2)y d\theta = \lambda I_0(c) - \mu I_1(c), \qquad (3.23)$$

where $I_0(c) = \int\int dy d\theta$ is the area of the region enclosed by the curve $F(\theta, y) = c$ and $I_1(c)$ is the trice inertia momentum of this region with respect to the horizontal axis. It was shown in [13] that the function $c \longmapsto I_1(c)/I_0(c)$ is monotone, i.e., that the equation $\lambda I_0(c) - \mu I_1(c) = 0$ has at most one solution. This means that system (3.20) for small λ and μ can have at most one limit cycle.

Here at the moment when the divergence of the field (3.20) at the focus C is zero, the Andronov–Hopf bifurcation takes place. One can check that it is not degenerate, $c_3 \neq 0$; readers are encouraged to do this. Let us provide corresponding calculations.

We put $y = 1 + z$, with small z, and expand the function (3.22) at $z = \theta = 0$. We get

$$F = -\frac{2}{3} + z^2 + \frac{1}{2}\theta^2 + \frac{1}{3}z^3 + \frac{1}{2}\theta^2 z + \cdots.$$

The famous Morse lemma allows to rewrite it as follows:

$$F + \frac{3}{2} = Z^2 + \frac{1}{2}\Theta^2,$$

where

$$Z \approx z\sqrt{1 + z/3} \approx z + \frac{1}{6}z^2, \quad \Theta \approx \theta\sqrt{1 + z} \approx \theta + \frac{1}{2}\theta z.$$

The inverse change is follows:

$$z \approx Z - \frac{1}{6}Z^2, \quad \theta \approx \Theta - \frac{1}{2}\Theta Z.$$

Thus,

$$y = 1 + Z - Z^2/6 + \cdots, \quad dy \wedge d\theta = (1 - 5Z/6 + \cdots)\, dZ \wedge d\Theta.$$

We rewrite $\lambda - 3\mu y^2$ as

$$\lambda - \mu - 3\mu\left(y^2 - 1\right) = \nu - 3\mu\left(2Z + 2Z^2/3 + \cdots\right),$$

where $\nu = \lambda - \mu$ is a new parameter (replacing λ). Then integral (3.23) takes the form

$$I\left(c\right) = \nu I_0\left(c\right) - \mu I_2(c),$$

where

$$I_0\left(c\right) = \int\!\!\int_{F \leq c} dz \wedge d\theta \approx \int\!\!\int dZ \wedge d\Theta,$$

$$I_1\left(c\right) = 3\int\!\!\int_{F \leq c} \left(y^2 - 1\right) dz \wedge d\theta \approx 3\int\!\!\int \left(2Z - Z^2\right) dZ \wedge d\Theta.$$

Using the polar type variables r, φ such that $Z = r \cos \varphi$ and $\Theta = \sqrt{2} r \sin \varphi$ and denoting $\tilde{c} = c + \frac{2}{3}$, we get

$$I_0 \approx \sqrt{2\pi} \tilde{c}, \quad I_2 \approx -\frac{3}{4} \sqrt{2\pi} \tilde{c}^2$$

as $\tilde{c} \to 0^+$. Therefore,

$$I(c) \approx \sqrt{2\pi} \left(\nu \tilde{c} + \frac{\mu}{4} \tilde{c}^2 \right).$$

It is exactly like in the Andronov–Hopf bifurcation. The parameter ν corresponds to the real part of the complex eigenvalues and μ is proportional to the focus quantity c_3.

3.7.   Codimension one non-local bifurcations

3.7.1.   *Limit cycle bifurcations*

Let $\gamma \subset M$ be a closed phase curve of a vector field $v_0(x)$ on a manifold M (n-dimensional). Additionally, the field $v_0(x)$ is included into a 1-parameter family $v_\mu(x)$, $\mu \in (\mathbb{R}, 0)$, of vector fields on M. Take a section $S \subset M$ transversal to γ. For μ close to 0, we get a family of Poincaré return maps $P_\mu : S \longmapsto S$. By identifying S with $(\mathbb{R}^{n-1}, 0)$, we get a family of local transformations

$$f_\mu : \left(\mathbb{R}^{n-1}, 0\right) \longmapsto \left(\mathbb{R}^{n-1}, 0\right), \quad f_\mu(z) = f(z; \mu),$$

such that $f_0(0) = 0$. So,

$$f_0(z) = Az + \cdots.$$

We also assume that for $\mu = 0$ the orbit γ is non-hyperbolic, i.e., the fixed point $z = 0$ of the diffeomorphism f_0 is non-hyperbolic. Depending on the type of the non-hyperbolicity, we have different types of bifurcations. Here we will discuss only two of them.

A. Saddle-node bifurcation for a limit cycle. We have $\lambda_1 = 1$ and $|\lambda_j| \neq 1$ ($j > 1$) for the eigenvalues of the matrix A. Reducing the situation to the center manifold (i.e., one dimension for

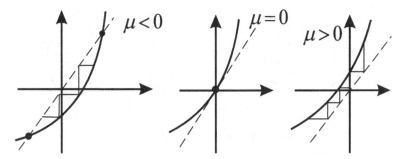

Fig. 3.15. Saddle-node bifurcation for diffeomorphisms.

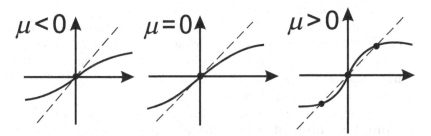

Fig. 3.16. Period doubling bifurcation for diffeomorphisms.

diffeomorphism and one for parameter) and applying appropriate conditions of non-degeneracy (i.e., $f''_{zz}(0,0)f'_\mu(0,0) \neq 0$), we reduce the situation to the following model family:

$$f(z;\mu) = z + \mu \pm z^2, \quad (z,\mu) \in (\mathbb{R}^2, 0). \qquad (3.24)$$

The appropriate bifurcations are shown in Fig. 3.15. We see that for $\mu < 0$ we have two limit cycles that merge at $\mu = 0$ and then disappear for $\mu > 0$.

B. **Period doubling bifurcation.** We have $\lambda_1 = -1$ and $|\lambda_j| \neq 1$ for $j > 1$. Because the return map changes the orientation, the center manifold for the orbit is a Möbius band (see Fig. 3.16). The model family of maps in this case is

$$f_\mu(z) = f(z,\mu) = (-1 + \mu)z \pm z^3.$$

Of course, for any μ the above transformation has only one fixed point, $z = 0$. But its second iteration has the form

$$f_\mu \circ f_\mu(z) = (1 - 2\mu)z \mp 2z^3 + \cdots \qquad (3.25)$$

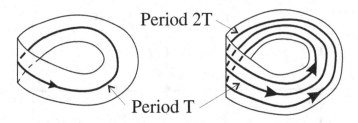

Fig. 3.17.   Period doubling bifurcation for limit cycles.

and has two additional fixed points $z_{1,2} \approx \sqrt{\mp\mu}$ for $\mp\mu > 0$. These fixed points correspond to the periodic orbit for f_μ with period 2. From this comes the bifurcation's name; sometimes also the name *pitchfork bifurcation* is used (from the shape of the curve of periodic points on the plane (μ, z)). Like in the case of Andronov–Hopf bifurcation the pitchfork bifurcations are divided into *subcritical* (i.e., unsafe), when $\mp = +$ in Eq. (3.25) and *supercritical* (i.e., safe), when $\mp = -$.

The appropriate bifurcations for $\{f_\mu\}$ are shown in Fig. 3.17.[e]

C. **Complex multipliers without strong resonance.** We assume that the return map is two-dimensional, i.e., the periodic trajectory γ of period T for v_0 has a neighborhood in the form of a full torus $\gamma \times \mathbb{D}^2$, $\mathbb{D}^2 = (\mathbb{C}, 0)$.

Using a complex variable $z \in (\mathbb{C}, 0)$, the return map takes the form

$$f_0(z, \bar{z}) = \lambda z + \cdots, \quad \lambda = e^{2\pi i \alpha}.$$

By assumption, either $\alpha \notin \mathbb{Q}$ (no resonance) or $\alpha = \frac{p}{q} \in \mathbb{Q}$, with $\gcd(p, q) = 1$ and $q \geq 5$ (weak resonance).

By a version of the Poincaré–Dulac normal form for diffeomorphisms, the map f_0 can be reduced to

$$z \mapsto \lambda z + cz |z|^2 + O(|z|^4).$$

---

[e]The period doubling bifurcation lies at the basis of a known Feigenbaum bifurcation for some irreversible transformations $g : I \longmapsto I$ of an interval. First, a fixed point loses stability when passing an eigenvalue by -1. Then the arising periodic period of period 2 again loses stability and a periodic orbit of period 2^2 is created, etc. For the parameter's limiting value, we have the Feigenbaum bifurcation.

Indeed, the terms $z^k \bar{z}^l$ in the expansion of f_0 correspond to resonances $\lambda = \lambda^{k-l}$ (see Remark 3.23). We impose the genericity assumption

$$\mathrm{Re}\, c \neq 0.$$

The generic deformation of the map f_0 is as follows:

$$f_\mu (z, \bar{z}) = \lambda(\mu)\, z + c(\mu)\, z\, |z|^2 + O(|z|^4), \quad |\lambda(\mu)| = 1 + \mu.$$

One can show that there exists an invariant closed curve δ_μ, close to the circle $\{|z| = \sqrt{-\mu/\mathrm{Re}\, c}\}$, when $\mu/\mathrm{Re}\, c < 0$; this corresponds to an invariant torus for a corresponding vector field v_μ. It is a version of the Andronov–Hopf bifurcation for diffeomorphisms. Moreover, this invariant circle δ_μ is normally hyperbolic, i.e., either attracting or repelling. But the dynamics on this circle, i.e., $f_\mu|_{\delta_\mu}$, is not determined. We only know that it is close to a rotation by the angle $2\pi\alpha$. There can exist periodic points of period q, when $\alpha = p/q$ (corresponding to periodic solutions of v_μ of period $\sim qT$). There can be no periodic points or there can exist periodic points of very large period in the case $\alpha \notin \mathbb{Q}$.

3.7.2. *Separatrix connection bifurcations*

There are two such bifurcations:

A. **Connection of two saddles.** This case is easy.
Before perturbation we have the situation like in the left phase portrait in Fig. 3.18. Any versal deformation v_μ is such that the separatrix connection splits up. The parameter μ is the distance between the right unstable separatrix of the left saddle and the left stable separatrix of the right saddle. This distance is measured at a section S transversal to the separatrix connection. We do not write formulas for v_μ and we do not draw pictures.

B. **Separatrix loop.** This case is slightly more complicated.
Before deformation we have the right phase portrait from Fig. 3.18. Like in the previous case, the small parameter μ of a versal deformation v_μ is the distance between the left unstable and left stable separatrices of the saddle. But we need the

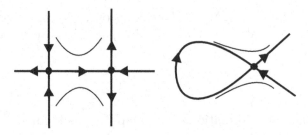

Fig. 3.18. Separatrix connection and separatrix loop.

following non-degeneracy assumption:

$$\kappa = |\lambda_2/\lambda_1| \neq 1,$$

where $\lambda_1 > 0$ and $\lambda_2 < 0$ are the eigenvalues of the saddle.

To analyze the phase portraits, we firstly choose local coordinates near the saddle such that the position of the saddle as well as its separatrices do not change with variation of the parameter. So, we assume the local system

$$\dot{x} = x\,(\lambda_1 + \cdots), \quad \dot{y} = y\,(\lambda_2 + \cdots),$$

where the eigenvalues $\lambda_j = \lambda_j(\mu)$ can depend on μ. Assume that the local parts of the separatrix loop are $W_+^u = \{(x,0) : x \geq 0\}$ and $W_+^s = \{(0,y) : y \geq 0\}$. Define also local sections: $S_1 = \{(x,a)\}$, transversal to W_+^s at $(0,a)$, $a > 0$, and $S_2 = \{(a,y)\}$, transversal to W_+^u at $(a,0)$.

Next, we define a Poincaré type return map $f : S_1 \mapsto S_1$ as compositions of two maps $f_1 : S_1 \mapsto S_2$ and $f_2 : S_2 \mapsto S_1$.

The map f_1 is called the Dulac map and takes the form

$$x \mapsto y = bx^\kappa + \cdots, \quad x \geq 0, \tag{3.26}$$

where $b = b(\mu) > 0$ and $\kappa = \kappa(\mu)$ is the minus ratio of eigenvalues (Exercise 3.54). The map f_2 is regular and can be written as follows:

$$y \mapsto x = \mu + cy + \cdots,$$

where $c = c(\mu) > 0$. Their composition takes the form

$$f(x) = \mu + dx^\kappa + \cdots, \quad x \geq 0, \tag{3.27}$$

where $d = d(\mu) > 0$.

The fixed "positive" points of this map, i.e., such that $x > 0$, correspond to limit cycles of the vector field v_μ.

It is easy that, if $\kappa > 1$, then for $\mu > 0$ the map (3.27) has a positive fixed points near $x = 0$, which corresponds to a stable limit cycle and for $\mu < 0$ we have only splitting of the separatrices. If $\kappa < 1$ then an unstable limit cycle bifurcates from the loop for $\mu < 0$.

3.8. Codimension two bifurcations

The codimension two bifurcations can be divided into three groups.

In the first group, we have easy generalizations of some codimension one bifurcations: codimension two saddle-node bifurcation (Exercise 3.55), codimension two Hopf bifurcation (Exercise 3.56), codimension two saddle-node bifurcation for diffeomorphisms (Exercise 3.57) and codimension two period doubling bifurcation (Exercise 3.58).

Into the second group we include non-local bifurcations of higher dimensional dynamical systems, which are highly complicated and which are not studied in this book.

The bifurcations from the third group are treated briefly as follows.

3.8.1. *Bogdanov–Takens bifurcation*

This is the case when the both eigenvalues of the linearization matrix A of a singular point of a plane vector filed vanish but A is non-zero. Thus, the codimension is two and A is nilpotent. We can assume

$$\dot{x} = y + \cdots, \quad \dot{y} = \cdots.$$

We want to find some normal form for this singularity. Of course, we cannot use the Poincaré–Dulac theorem. Nevertheless, we have the following result.

Theorem 3.32 (Takens). *There exists a formal change of variables leading to the following Liénard type vector field:*

$$\dot{x} = y, \quad \dot{y} = a(x) + b(x)y. \tag{3.28}$$

Proof. We have the homological operator $\mathrm{ad}_{y\partial/x}$ acting on vector fields $Z = f\partial/\partial x + g\partial/\partial y$ as follows:

$$Z \longmapsto (yf_x - g)\,\partial/\partial x + yg_x\partial/\partial y$$

(cf. the proof of Theorem 3.19 and Exercise 3.43). We claim that the subspace $\{(a(x) + b(x)y)\,\partial/y\}$ is complementary to the image of this operator.

Note that any series in x, y is of the form $yf_x - g$. But, when $yf_x - g = 0$, then $yg_x = y^2 f_{xx}$, where f_{xx} is any series in x, y. $\square$

So, we assume the following form of v_0:

$$\dot{x} = y, \quad \dot{y} = \alpha x^2 + O\left(x^3\right) + \left(\beta x + O\left(x^2\right)\right)y.$$

The genericity assumption is

$$\alpha \neq 0, \quad \beta \neq 0.$$

By a suitable normalization (maybe with reversion of the time) we can assume $\alpha = -1$ and $\beta = 1$. A natural 2-parameter deformation relies upon adding $\mu_1 + \mu_2 y$ to $\dot{y}$; thus the model family is following:

$$\dot{x} = y, \quad \dot{y} = \mu_1 + \mu_2 y - x^2 + xy. \tag{3.29}$$

The analysis of the phase portraits of his system begins with the singular points. There are two of them $(x_{1,2}, 0) = \left(\pm\sqrt{\mu_1}, 0\right)$ with the linearization matrices $\begin{pmatrix} 0 & 1 \\ -2x_{1,2} & \mu_2 + x_{1,2} \end{pmatrix}$, when $\mu_1 = 0$; then the line $\{\mu_1 = 0\}$ is the line of saddle-node bifurcation. The point $(x_2, 0)$ is a saddle and the point $(x_1, 0)$ is either a node or a focus.

When $\mu_2 + x_2 = \mu_2 + \sqrt{\mu_1} = 0$ the points $(x_1, 0)$ becomes a weak focus and an Andronov–Hopf bifurcation takes place. The creation of a limit cycle and its properties away from the Andronov–Hopf bifurcation are the only delicate points in this case.

To study this limit cycle, we introduce the following normalization:

$$x = \mu_1^{1/2}X, \quad y = \mu_1^{3/4}Y, \quad dt = \mu^{-1/4}dT.$$

Then we get the system

$$\frac{dX}{dT} = Y, \quad \frac{dY}{dT} = 1 - X^2 + \nu_1 Y + \nu_2 XY,$$

where $\nu_1 = \mu_2\mu_1^{-1/4}$ and $\nu_2 = \mu_1^{1/4}$ are small parameters. The latter system is a perturbation of a Hamiltonian system with the Hamilton function

$$H = \frac{1}{2}Y^2 - X + \frac{1}{3}X^3.$$

Accordingly to the averaging method (developed in Section 4.1 in the next chapter), the approximate condition for a limit cycle bifurcating from an oval $\gamma_h \subset \{H = h\}$ is the vanishing of the following Abelian integral:

$$I(h) = \oint_{\gamma_h} (\nu_1 + \nu_2 X) Y \, dX.$$

The latter integral was firstly studied by R. Bogdanov (see also Example 4.9 and Exercise 4.35 in the next chapter). The function $I(h)$ has at most one zero, which implies that the unique limit cycle evolves from the point $(X, Y) = (1, 0)$ for $\nu_1 = -\nu_2$ to a separatrix loop of the saddle $(X, Y) = (-1, 0)$ for $\nu_1 = -\frac{5}{7}\nu_2$. The corresponding bifurcational curves in the μ-plane are $H = \{\mu_2 \approx -\sqrt{\mu_1}\}$ and $L = \{\mu_2 \approx -\frac{5}{7}\sqrt{\mu_1}\}$.

In Fig. 3.19, we present the bifurcational diagrams and corresponding phase portraits for the family (3.29).

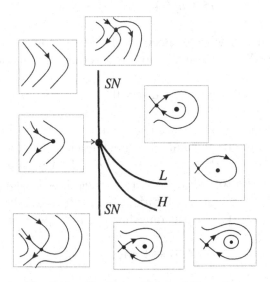

Fig. 3.19.   Bogdanov–Takens bifurcation.

3.8.2.   Saddle-node–Hopf bifurcation

Before perturbation we have a three-dimensional vector field v_0 near a singular point with eigenvalues $0, \pm i\omega$, i.e.,

$$\dot{x}_1 = \cdots, \quad \dot{x}_2 = -\omega x_3 + \cdots, \quad \dot{x}_3 = \omega x_2 + \cdots.$$

The phase portrait of the linear system is the foliation into the circles $\{x_1 = c,\ x_2^2 + x_3^2 = r^2\}$. So, it is natural to consider corresponding Poincaré type return map to the section $S = \{x_3 = 0,\ x_2 \geq 0\}$. Since the section is two-dimensional, the analysis of this map is not easy. To simplify the problem one applies the following approximation.

This approximation relies upon averaging of the vector field v_0 over the curves of the latter foliation. For example, if $\dot{x}_1 = f(x_1, x_2, x_3)$, then we take the averaged velocity of x_1, i.e.,

$$\dot{x}_1 = \tilde{f}(x_1, r) = \frac{1}{2\pi} \int_0^{2\pi} f(x_1, r\cos\varphi, r\sin\varphi)\, d\varphi.$$

After averaging and denoting $x = x_1$, $y = r$, we get the following two-dimensional autonomous system:

$$\dot{x} = ax^2 + by^2 + \cdots, \quad \dot{y} = cxy + \cdots,$$

which is invariant with respect to the involution $y \longmapsto -y$ ($\mathbb{Z}_2$ – equivariance). We impose the non-degeneracy assumption

$$abc \neq 0.$$

By suitable normalization, we can assume $b = \pm 1$ and $c = -2$.

We also need to normalize some higher order term, the cubic ones: $(dx^3 + exy^2)\, \partial_x + (fx^2y + gy^3)\, \partial_y$. To this aim, we use the homological operator ad_{V_0}, $V_0 = (ax^2 + by^2)\, \partial_x + cxy\partial_y$, applied to the vector fields $Z = (\alpha x^2 + \beta y^2)\, \partial_x + \gamma xy\partial_y$, and the time change $dt \longmapsto dt/(1 + \delta x)$. It turns out that the full operator

$$(Z, \delta) \longmapsto \mathrm{ad}_{V_0} + \delta x V_0 \tag{3.30}$$

has rank 3 and the $x^3 \partial_x$ lies outside its image (Exercise 3.59).

The deformation relies upon adding the vector field $(\mu_1 + \mu_2 x)\,\partial_x$. Therefore, we have the following model families:

$$x = \mu_1 + \mu_2 x + ax^2 \pm y^2 + dx^3, \quad y = -2xy \qquad (3.31)$$

with the non-degeneracy condition

$$d \neq 0.$$

There are at most two singular points $p_{1,2}$ on the invariant line $y = 0$; we skip analysis of their bifurcations. There are also two symmetric singular points $q_{1,2} = (0, \pm\sqrt{\mp\mu_1})$ if $\mp\mu_1 > 0$.

Consider the point $q_1 = (0, y_0)$ with the characteristic polynomial $P(\lambda) = \lambda^2 - \mu_2\lambda + 4cy_0^2$, $c = \pm 1$. In the case $c = 1$, we have the Andronov–Hopf bifurcation for $\mu_2 = 0$.

To study this bifurcation and global behavior of the corresponding limit cycle we apply the following normalization:

$$x \longmapsto X = x/\sqrt{|\mu_1|}, \quad y \longmapsto Y = y/\sqrt{|\mu_1|}, \quad dt \longmapsto \sqrt{|\mu_1|}dt.$$

Then we get the system

$$\dot{X} = -1 + aX^2 + Y^2 + \nu_1 X + \nu_2 X^3, \quad \dot{Y} = -2XY,$$

where $\nu_1 = \mu_2/\sqrt{|\mu_1|}$ and $\nu_2 = \sqrt{|\mu_1|}$ are small parameters.

For $\nu_1 = \nu_2 = 0$ this system has the integrating factor Y^{a-1} (here $Y > 0$) and the first integral (see Fig. 3.20)

$$H = Y^a \left(X^2 + Y^2/(a+2) - 1/a \right).$$

Like in other similar situations, the problem of limit cycles reduces to the problem of zeroes of the following Melnikov function:

$$I(h) = \oint_{H=h} Y^{a-1} \left(\nu_1 X + \nu_2 X^3 \right) dY.$$

It turns out that the latter function has at most one zero (see [14]).

Using this and the easy bifurcations of the singular points, one can draw the bifurcation diagrams and phase portraits (like in Fig. 3.19). Of course, the cases with $c = 1$ and $c = -1$ as well as with $a > 0$ and $a < 0$ are qualitatively different. Nevertheless, we do not do it here and we refer to the monographs [14,15].

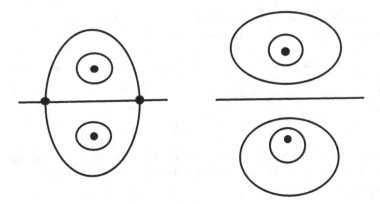

Fig. 3.20.   The first integral.

We note only that in the situation with $c = 1$ and $a > 0$ the limit cycle is born at p_1 and grows to a double separatrix connection of two saddles at the invariant line $y = 0$.

Finally, we recall that system (3.31) is a result of averaging of a three-dimensional vector field v_μ. Therefore, the singular points $q_{1,2} = (0, \pm\sqrt{|\mu_1|})$ outside the invariant line correspond to a closed phase curves of v_μ. The pitchfork bifurcation when the symmetric points q_1 and q_2 tend to a singular point p_j on the invariant line corresponds to an Andronov–Hopf bifurcation on a center manifold of a corresponding singularity of v_μ.

Next, the limit cycle born in the Andronov–Hopf bifurcation from q_1 corresponds to an invariant torus for v_μ; it usually supports a non-trivial dynamics. But we cannot say much about the corresponding bifurcation from the limit cycle.

When the cycle becomes a double separatrix connection, for the vector field v_μ, the corresponding symmetric double connection would consist of a segment in the symmetry axis between two hyperbolic singular points, say q_1 and q_2, and a surface (approximately a 2-sphere) being a simultaneous invariant manifold of the points $q_{1,2}$. When v_μ is not perfectly symmetric, i.e., with respect to rotations in the x_2x_3-plane, the invariant manifolds of q_1 and of q_2 can intersect transversally along several phase curves and this rather inevitably leads to a chaotic dynamics. Also the connection along the symmetry axis would be destroyed.

3.8.3. Hopf–Hopf bifurcation

Before perturbation we have a singular point of a four-dimensional vector field with imaginary eigenvalues without resonance, i.e.,

$$\dot{x}_1 = -\omega_1 x_2 + \cdots, \quad \dot{x}_2 = -\omega_1 x_1 + \cdots,$$

$$\dot{x}_3 = -\omega_2 x_4 + \cdots, \quad \dot{x}_4 = \omega_2 x_3 + \cdots,$$

where $\omega_1/\omega_2 \notin \mathbb{Q}$.

The phase curves of the corresponding linear vector field lie in the 2-tori $\{x_1^2 + x_2^2 = r_1^2, \, x_3^2 + x_4^2 = r_2^2\}$, parametrized by angles φ_1 and φ_2, and are dense in these tori. In the case of general system, it is natural to apply the averaging procedure, i.e., replacing $f(x)$ with

$$\tilde{f}(r_1, r_2) = (2\pi)^{-2} \int \int f(x) \, \mathrm{d}\varphi_1 \mathrm{d}\varphi_2$$

(see Section 4.2 in the next chapter). As a result, we get the following *amplitude system*:

$$\dot{x} = x(ax + by + \cdots), \quad \dot{y} = y(cx + dy + \cdots),$$

where $x = r_1^2 \geq 0$, $y = r_2^2 \geq 0$. A natural non-degeneracy condition is

$$abcd\,(ad - bc) \neq 0.$$

By applying a normalization, we can assume $a = b = 1$, $c = -\frac{\alpha+1}{\beta}$, $d = -\frac{\alpha}{\beta+1}$. Like in the saddle-node–Hopf bifurcation we want to reduce cubic terms using vector fields $Z = x(\gamma x + \delta y)\,\partial_x + (\varepsilon x + \zeta y)$, in the homological operator ad_{V_0}, $V_0 = x(ax + by)\,\partial_x + y(cx + dy)\,\partial_y$, and the time change $\mathrm{d}t \longmapsto \mathrm{d}t/(1 + \eta x + \theta y)$. The corresponding linear operator

$$(Z, \eta, \theta) \longmapsto \mathrm{ad}_{V_0} Z + (\eta v + \theta y) V_0 \qquad (3.32)$$

has rank 5 and the vector field $yR(x, y)\,\partial_y$, $R = x^2/\beta + 2xy/(\beta + 1) + y^2/(\beta + 2)$, lies outside its image (Exercise 3.60). Thus, we have the model family

$$\dot{x} = x(\mu_1 + x + y), \; \dot{y} = y\left(\mu_2 - \frac{\alpha+1}{\beta}x - \frac{\alpha}{\beta+1}y + R(x,y)\right),$$
$$(3.33)$$

where we have included the perturbation (with the small parameters μ_1 and μ_2).

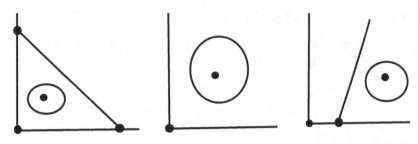

Fig. 3.21.   The first integral.

This system has the obvious singular point $p_0 = (0,0)$. There are also singular points p_1 and p_2 on the invariant axes $x = 0$ and $y = 0$, respectively. Finally, there exists a singular point p_3 outside the invariant axes; there an Andronov–Hopf bifurcation can take place.

Without going into details (which can be found in [14,15]) one performs an normalization leading to the system

$$\dot{X} = X\left(\pm 1 + X + Y\right),$$

$$\dot{Y} = Y\left(\mp\frac{\alpha}{\beta} - \frac{\alpha+1}{\beta}X - \frac{\alpha}{\beta+1}Y + \nu_1 + \nu_2 R\left(X,Y\right)\right)$$

with small parameters ν_1 and ν_2. For $\nu_1 = \nu_2 = 0$, the latter system has the integrating multiplier $X^{\alpha-1}Y^{\beta-1}$ and first integral

$$H = X^{\alpha}Y^{\beta}\left(\pm\frac{1}{\beta} + \frac{X}{\beta} + \frac{Y}{\beta+1}\right)$$

(cf. Example 2.38 from the previous chapter and Fig. 3.21). As usual, the problem of limit cycles here boils down to the problem of zeroes of the following Melnikov function:

$$I\left(h\right) = \oint_{H=h} X^{\alpha-1}Y^{\beta}\left(\nu_1 + \nu_2 R\left(X,Y\right)\right)\mathrm{d}X.$$

It was proved that this function has at most one zero.

Now, the bifurcational diagrams and corresponding phase portraits can be drawn. But there are many cases and we skip this.

Finally, we make some comments about the four-dimensional systems v_μ. The singular points p_1 and p_2 on the axes correspond to limit cycles of v_μ. The singular point p_3 corresponds to and invariant 2-torus with dynamics close to rotations with frequencies ω_1 and ω_2, but not determined. Finally, the limit cycle born in the Andronov–Hopf bifurcation corresponds to an invariant 3-torus, also with non-determined dynamics.

When p_1 tends to p_0 (or p_2 tends to p_0), then we have the usual Andronov–Hopf bifurcation for v_μ. The cases when p_3 tends to p_1 (or to p_2) correspond to limit cycle bifurcations without strong resonance (because ω_1/ω_2 is irrational).

The bifurcation from a 2-torus to a 3-torus is quite new and weakly determined.

Next, the limit cycle for the two-dimensional system can grow to a triangle composed of separatrices of three saddles. In the three-dimensional situation, the corresponding connections are destroyed with creation of a chaotic dynamics.

3.8.4. *Limit cycle bifurcations with strong resonances*

Recall that these are perturbations of a vector field v_0 with a limit cycle γ such that its return map has two-dimensional center manifold and a pair of eigenvalues of the fixed point lies at the unit circle. We have $\lambda_{1,2} = e^{\pm 2\pi i \alpha}$ with $\alpha = \frac{p}{q} \in \mathbb{Q}$ such that

$$q = 1, 2, 3, 4.$$

Also in these cases one applies an approximation using some averaging. Namely, one assumes that the dominating part of v_0 near γ is such that the corresponding phase curves form the so-called *Seifert foliation*. This means that, in the variables $(\theta, z) \in \mathbb{S}^1 \times (\mathbb{C}, 0)$, such that $\gamma = \mathbb{S}^1 \times \{0\}$, the dominating part of the vector field is

$$\dot{\theta} = 1, \quad \dot{z} = i\alpha z$$

with the phase curves $\{z = z_0 e^{i\alpha\theta}\}$.

Then, if $\dot{z} = f_0(z, \bar{z})$ for v_0 then the averaging along the leaves of the Seifert foliation yields the following equation:

$$\dot{z} = w_0(z, \bar{z}) = \frac{1}{2q\pi} \int_0^{2q\pi} f_0\left(ze^{i\theta}, ze^{-i\theta}\right) d\theta.$$

The resulting vector field is $\mathbb{Z}_q$-equivariant (invariant with respect to the rotation $z \longmapsto e^{2\pi i/q} z$).

All the above four cases are treated separately as follows.

A. $q = 1$. Here $\lambda_1 = \lambda_2 = 1$. The corresponding Seifert foliation is trivial and the averaging and introduction of deformation
leads to the Bogdanov–Takens bifurcation (3.29). So, we skip this
analysis.

We note only, as the time one flow of the vector field (3.29)
is only approximation of the return map, the dynamics on the
invariant torus (corresponding to the limit cycle for (3.29)) can
be non-trivial and the situation with the saddle loop for (3.29)
can lead to a chaotic dynamics for the return map.

B. $q = 2$. Thus, $\lambda_1 = \lambda_2 = -1$. Here the averaging leads to vector fields equivariant with respect to the involution $(x, y) \longmapsto$
$(-x, -y)$.

We have (after application of the Takens Theorem 3.32)

$$\dot{x} = y, \quad \dot{y} = \mu_1 x + \mu_2 y + ax^3 + bx^2 y, \quad ab \neq 0, \qquad (3.34)$$

where we can assume $a = \pm 1$ and $b = 1$ (by a normalization).

In general, we have at most three singular points $p_0 = (0, 0)$
and $p_{1,2} = (\pm x_{1,2}, 0)$, whose bifurcations are easy.

The Andronov–Hopf bifurcation takes place at p_0, when $a = 1$
and $\mu_1 < 0$, and the symmetric points p_1 and p_2, when $a = -1$
and $\mu_1 > 0$. After suitable normalization we get the following
perturbation of a Hamiltonian vector field:

$$\dot{X} = Y, \quad \dot{Y} = \pm \left(X^3 - X \right) + \nu_1 X + \nu_2 X^2 Y$$

with the Hamilton function

$$H = \frac{1}{2} Y^2 \pm \left(\frac{1}{2} X^2 - \frac{1}{4} X^4 \right)$$

and small parameters $\nu_{1,2}$. The corresponding Abelian integral

$$I(h) = \oint \left(\nu_1 + \nu_2 X^2 \right) Y \, dX$$

along an oval (compact component) $\gamma_h \subset \{H = h\}$ has at most
one zero. (Note that for $a = -1$ we have two symmetric ovals
around $(\pm 1, 0)$.)

From this the bifurcational diagram and phase portraits can
be drawn.

C. $q = 3$, i.e., $\lambda_{1,2} = e^{\pm 2\pi i/3}$. The corresponding deformation is as follows:

$$\dot{z} = \mu z + Az\,|z|^2 + B\bar{z}^2, \quad AB \neq 0,$$

where $\mu = \mu_1 + i\mu_2$ is small and we can assume $B = 1$ and $A = a + ib$.

Note that the quadratic term $\bar{z}^2$ is dominating over the term $Az\,|z|^2$. So, the approximate equation for the singular points takes the form $\mu z = -\bar{z}^2$, with one solution $z_0 = 0$ and three symmetric solutions $z_{1,2,3}$. Their bifurcations are also $\mathbb{Z}_3$-symmetric, i.e., they can collide only with z_0.

The Andronov–Hopf bifurcation concerns only z_0 and takes place for $\mu_1 = 0$, i.e., $\mu = i\mu_2$. After suitable normalization, we get the system

$$\dot{Z} = iZ + \bar{Z}^2 + \nu_1 Z + \nu_2 Z\,|Z|^2,$$

which is a perturbation of a Hamiltonian system with the Hamilton function (see Exercise 2.58)

$$H = \frac{1}{2}|Z|^2 - \frac{i}{6}(Z^3 - \bar{Z}^3);$$

note also that the corresponding quadratic system belongs to three components Q_3^H, Q_3^R and Q_3^{LV} (see Example 2.16) and is called the Hamiltonian triangle. The corresponding Abelian integral

$$I(h) = \frac{1}{2i} \oint_{H=h} (\nu_1 + \nu_2\,|Z|^2)\,(Z d\bar{Z} - \bar{Z} dZ)$$

$$= -\oint (\nu_1 + \nu_2 r^2)\,r^2 d\varphi,$$

where $Z = re^{i\varphi}$, has at most one zero.

Again, we skip drawing bifurcational diagrams and phase portraits.

D. $q = 4$. Thus, the eigenvalues are $\pm i$ and considered phase portraits are invariant with respect to rotation by the angle $\frac{\pi}{2}$.

The model family is as follows:

$$\dot{z} = \mu z + Az\,|z|^2 + B\bar{z}^3, \quad B \neq 0,$$

where $\mu = \mu_1 + i\mu_2$ is small, $A = a + ib$ and we can assume $B = 1$. Note that we have two cubic terms in the right-hand side and this means that the situation is highly complicated (and not completely investigated).

There can exist nine singular points (including $z = 0$) which are subject to many bifurcations. There are limit cycles (not fully investigated). There are separatrix connections and separatrix loops. The known non-degeneracy conditions are as follows:

$$|A| \neq 1, \quad |a| \neq 1, \quad |b| \neq \left(1 + a^2\right)/\sqrt{1 - a^2}$$

(see [6,15]).

Here we finish our discussion of the bifurcations of vector fields. For more details on the bifurcation described above and other bifurcations, we refer the readers to [6,10,12–15].

3.9. Exercises

Exercise 3.33.

(a) Show the openness of the transversality property in case (2), i.e., $m > n - k$, in the proof of Theorem 3.6. Recall that this property means $\text{rk}\,C = n - k$.

Hint: Use the Implicit Function Theorem to show that, if $f_0 \pitchfork B$ then $f_0^{-1}(B) \subset A$ in a neighborhood of $x_0 \in f_0^{-1}(B)$ is a submanifold in A of codimension $n - k$, locally defined as $x_1 = \cdots = x_{n-k} = 0$. Then show that for any f near to f_0, also $f^{-1}(B)$ is a local submanifold of codimension $n - k$ close to $f_0^{-1}(B)$. It is useful to consider the map $\mathcal{F} : (A, x_0) \times C^r(A, M) \mapsto \mathbb{R}^{n-k}$, $(x, f) \longmapsto \mathcal{F}(x, f) = (f_1(x), \ldots, f_{n-k}(x))$ and apply the Implicit Function Theorem with respect to the variables $x_1, \ldots, x_{n-k}$; here $\det \partial(f_1, \ldots, f_{n-k})/\partial(x_1, \ldots, x_{n-k}) \neq 0$.

Finally, deduce from the openness of the subset of matrices of the rank $n-k$ in the space of the matrices of dimension $(n-k) \times m$ that any f close to f_0 is transversal to B in a neighborhood of x_0.

(b) Show that $f \pitchfork B$ if and only if the value 0 is not critical for the mapping g.

(c) Complete the proof of Theorem 3.6 in the case of general manifolds.

   Hint: Use a corresponding partition of unity $1 = \sum \varphi_j(x)$ in A associated with local affine maps in A, M and B. Choose local perturbations in the form $f_\varepsilon = (g - \psi_k(x)z_k, h(x))$, with suitable functions $\psi_k(x)$ with a compact support and "small" vectors z_k. Finally, use the compactness of A.

Exercise 3.34. Prove directly the Sard theorem for cases $m < l$ and $m = l > 1$.

Exercise 3.35. Prove the properties (3.3) from the proof of the Sard theorem.

Exercise 3.36. Prove formula (3.8) for the recurrence in the proof of Theorem 3.16.

Exercise 3.37. Using reduction to the center manifold study the stability of the point $(0,0,0)$ of the vector field

$$\dot{x} = -x - 2y - z - y^3, \quad \dot{y} = 2x - 4y + z - z^3, \quad \dot{z} = -z - x^3.$$

Exercise 3.38. Find the approximation of the center manifold modulo cubic terms for the point $(0,0,0)$ of the system $\dot{x} = -y + x^2 + zy$, $\dot{y} = x + xy + z^2$, $\dot{z} = z + xy$.

   Compute also the focus quantity c_3 of the focus at the center manifold.

Exercise 3.39. Consider the family of planar vector fields $\dot{x} = -x + y^2 + \mu x$, $\dot{y} = -\sin x$, $(x, y, \mu) \in (\mathbb{R}^3, 0)$.

(a) Find the one-dimensional center manifold in $\mathbb{R}^2$ for $\mu = 0$.

(b) Find the two-dimensional center manifold in $\mathbb{R}^3$ and study the bifurcation taking place on it.

Exercise 3.40. Show that the inverse transformation to (3.11) has the form as in the proof of Theorem 3.19.

Exercise 3.41. Show that in any other case of the node than in Example 3.20, i.e., when $1 < \lambda_2/\lambda_1 \notin \mathbb{N}$, the Poincaré–Dulac normal form is linear (no nonlinear resonant terms).

Exercise 3.42. Generalize Example 3.22 for the case of $p : -q$ resonant saddle, i.e., when $0 < \lambda_1/\lambda_2 = -\frac{p}{q} \in \mathbb{Q}$ (reduced fraction).

Exercise 3.43. Show that the homological operator (3.14) from the proof of the Poincaré–Dulac theorem takes the form

$$Z \longmapsto \mathrm{ad}_{V_0} Z = [V_0, Z],$$

where $[V_0, Z]$ is the commutator of two vector fields from Section 6.5 in the Appendix (see also Exercise 6.63). It can be regarded as the first level homological operator, because it uses the first term in the expansion of the vector field $V = V_0 + \cdots$.

Exercise 3.44. Show that the normal form for singular vector fields of the type $\dot{x} = ax^{k+1} + \cdots$, $a \neq 0$, $k \geq 1$, is $\dot{x} = \pm x^{k+1} + \alpha x^{2k+1}$, $\alpha \in \mathbb{R}$.
Hint: Use the homological operator $Z = b(c) \frac{\partial}{\partial x} \longmapsto \mathrm{ad}_{V_0} Z$ with $V_0 = ax^{k+1} \frac{\partial}{\partial x}$.

Exercise 3.45. Assume that the first level normal for the saddle-node singularity form from Example 3.21 is $\left(a_{m+1} x_1^{m+1} + \cdots \right) \partial/\partial x_1 + x_2 \left(1 + b_1 x_1 + \cdots \right) \partial/\partial x_2$ with $a_{m+1} \neq 0$. Show that this form can be further reduced to

$$\left(\pm x_1^{m+1} + a x_1^{2m+1} \right) \partial/\partial x_1 + P_m(x_1) x_2 \partial/\partial x_2$$

or, equivalently to

$$P_m(x_1) \left\{ \left(\pm x_1^{m+1} + a x_1^{2m+1} \right) \partial/\partial x_1 + x_2 \partial/\partial x_2 \right\},$$

where P_m is a polynomial of degree $\leq m$ with $P(0) = 1$. Above $\left(\pm x_1^{m+1} + a x_1^{2m+1} \right) \partial/\partial x_1 + x_2 \partial/\partial x_2$ is the orbital normal form of the saddle-node singularity.
Hint: The vector fields $x_1^k \partial/\partial x_1$ and $x_1^k x_2 \partial/\partial x_2$ from the first level normal form span also the kernel of the first level homological operator ad_{V_0}, $V_0 = x_2 \partial/\partial x_2$. Use these terms in the second level homological operator ad_{V_1}, $V_1 = V_0 + a_{m+1} x_1^{m+1} \partial/\partial x_1$.

Exercise 3.46.

(a) Consider Example 3.22 with $V_0 = x_1 \partial/\partial x_1 - x_2 \partial/\partial x_2$, i.e., the $1 : -1$ resonant saddle. Assume that the first level normal form, in the variables $u = x_1 x_2$ and x_1 takes the form

$$\dot{u} = c_{m+1} u^{m+1} + \cdots, \quad \dot{x}_1 = x_1 \left(1 + a_1 u + \cdots \right),$$

where $c_{m+1} \neq 0$.
Show that it can be further reduced to

$$\dot{u} = P_m \left(u \right) \left(\pm u^{m+1} + cu^{2m+1} \right), \quad \dot{x}_1 = x_1 P_m(u),$$

corresponding to the vector field

$$Q_m \left(u \right) \left\{ V_0 + \left(\pm u^m + cu^{m+1} \right) E \right\},$$

where P_m and Q_m are polynomials of degree $\leq m$ and $E = x_1 \partial/\partial x_1 + x_2 \partial/\partial x_2$ is the Euler vector field.
Hint: Apply the homological operator ad_{V_1}, $V_1 = V_0 + \alpha u^m E$, to vector fields $u^k x_1 \partial/\partial x_1$ and $u^k x_2 \partial/\partial x_2$ spanning $\ker \mathrm{ad}_{V_0}$.
(b) Generalize this to the case of the $p : -q$ resonant saddle.

Exercise 3.47. Show that the local diffeomorphism $f(x) = \lambda x + \cdots$, with $\lambda = e^{2\pi i p/q}$, $q > 0$ and $\gcd(p, q) = 1$, can be reduced, using a formal conjugation $f \longmapsto h \circ f \circ h^{-1}$, to

$$\lambda x + a_{q+1} x^{q+1} + a_{2q+1} x^{2q+1} + \cdots.$$

Exercise 3.48. Show that condition (4) in Section 3.5 has the following interpretation. From condition (3), it follows that the equation $f'_x = 0$ has a unique solution $x = x_0(y, \mu)$ (local extremum point). Let $g(y, \mu) = f(x_0(y, \mu), y; \mu)$ (value at this point). Then $\frac{\partial f}{\partial \mu}(0, 0; 0) \neq 0$ $\iff \frac{\partial g}{\partial \mu}(0, 0) \neq 0$.

Exercise 3.49. Show that the coefficients $c_3, c_5, \ldots$ from point (4) of the assumptions to the Andronov–Hopf theorem coincide, up to positive constants, with the Poincaré–Lyapunov focus quantities from Definition 2.12. Find the relationship between the coefficients c_{2j+1}, d_{2j+1} and the coefficients a_j, b_j.

Exercise 3.50. Prove formula (3.17).
Hint: The reduction of finite number of resonant terms can be performed simultaneously for a 1-parameter family.

Exercise 3.51. Prove formula (3.18).

Exercise 3.52. Show rigorously that the equation $F = 0$ from the proof of Theorem 3.27 has exactly two solutions.

Exercise 3.53. Study the singular points of field (3.20) with $\lambda = 0$. Sketch the phase portraits for $\mu = 0$ and $\mu \neq 0$, but small.

Exercise 3.54. Prove that the Dulac map between sections to separatrices of a saddle with the absolute value of the ratio between eigenvalues κ takes the form (3.26).

Exercise 3.55. Study the bifurcations of equilibrium points for the 2-parameter families of one-dimensional vector fields, $\dot{x} = \lambda_1 + \lambda_2 x \pm x^3$ (deformation of a saddle-node singularity of codimension 2).

Exercise 3.56. Study the bifurcations of limit cycles for the families of two-dimensional vector fields $\dot{z} = (\lambda_1 + i) z + \lambda_2 z |z|^2 \pm z |z|^4$.

Exercise 3.57. Study the bifurcations of the fixed points for the families of one-dimensional diffeomorphisms $f_\lambda(x) = \lambda_1 + (1 + \lambda_2)x^2 \pm x^3$.

Exercise 3.58. Study the bifurcations of the periodic points for the families of diffeomorphisms $f_\lambda(x) = (-1 + \lambda_1) x + \lambda_3 x^3 \pm x^5$.

Exercise 3.59. Prove the statements about the operator (3.30) in the saddle-node–Hopf bifurcation.

Exercise 3.60. Prove the statements about the operator (3.32) in the Hopf–Hopf bifurcation.

Chapter 4

Equations with a Small Parameter

A small parameter in a differential equation can appear in essentially two ways: either on the right side or on the left side. In the first case, we deal with a small perturbation of a system, about which we know rather lot and in the second case there are the so-called relaxation oscillations. Both cases are discussed in the sequel sections.

4.1. Averaging

An example of a system from the first group is the known *van der Pol system*

$$\dot{x} = y, \quad \dot{y} = -x + \varepsilon(1 - x^2)y,$$

where ε is our small parameter. This is a special case of a **perturbation of Hamiltonian system with one degree of freedom** (with the Hamilton function $H = \frac{1}{2}(x^2 + y^2)$)

$$\dot{x} = \frac{\partial H}{\partial y} + \varepsilon P(x, y), \quad \dot{y} = -\frac{\partial H}{\partial x} + \varepsilon Q(x, y).$$

4.1.1. *Integrable Hamiltonian systems*

In applications often appear **Hamiltonian systems with many degrees of freedom** of the form

$$\dot{q}_i = \frac{\partial H}{\partial p_i}, \quad \dot{p}_i = -\frac{\partial H}{\partial q_i}, \quad i = 1, \ldots, n, \tag{4.1}$$

where q_i are generalized coordinates, p_i are generalized momenta and the function $H(q_1, \ldots, q_n, p_1, \ldots, p_n)$ is the **Hamiltonian function**, or the **Hamiltonian** (Exercises 6.59 and 6.60). In general, system (4.1) cannot be solved. However, there is a class of Hamiltonian systems fully solvable. For more informations about Hamiltonian systems, we refer the reader to Section 6.5 in Appendix and to [11].

Definition 4.1. System (4.1) is called **completely integrable** if there is a system of functionally independent first integrals $F_1 = H, F_2, \ldots, F_n$ such that each function F_j is the first integral for the Hamiltonian systems generated by other functions F_i, i.e., the Poisson brackets vanish:

$$\{F_i, F_j\} := \sum_{i=1}^{n} \frac{\partial F_i}{\partial q_i} \frac{\partial F_k}{\partial p_i} - \sum_{i=1}^{n} \frac{\partial F_i}{\partial p_i} \frac{\partial F_k}{\partial q_i} \equiv 0.$$

Examples of completely integrable systems are: the Kepler problem (see Example 4.4), some cases of the rigid body dynamics (see Example 4.5) and the geodesic flow on the surface of an ellipsoid (see [11]); the first two cases have two degrees of freedom and the ellipsoid can have any dimension.

Of course, in this case the motion lies on the sets $\{F_1 = c_1, \ldots, F_n = c_n\}$.

For systems which satisfy the condition of Definition 4.1 the following result holds, which we quote without proof (see [11] and Section 6.5 in Appendix).

Theorem 4.2 (Liouville–Arnold). *If the common levels $\{F_1 = c_1, \ldots, sF_n = c_n\}$ of a completely integrable Hamiltonian system are connected, compact and smooth, then they are tori $\mathbb{T}^n$.*

*Moreover, in a neighborhood of a given torus there exists a system of coordinates $(I_1, \ldots, I_n, \varphi_1, \ldots, \varphi_n)$, the so-called **action–angle variables**, in which system (4.1) takes the following Hamiltonian form:*

$$\dot{I}_j = 0, \quad \dot{\varphi}_j = \omega_j(I) = \partial H_0 / \partial I_j, \qquad j = 1, \ldots, n, \qquad (4.2)$$

where $H_0(I_1, \ldots, I_n) = H(q, p)$ is the Hamiltonian after the change. In general, the movement on the tori $\{I_1 = d_1, \ldots, I_n = d_n\}$, which

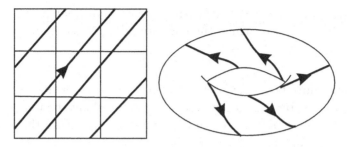

Fig. 4.1. Quasi-periodic dynamics on torus.

are parameterized by the angles φ_j mod 2π, *is periodic or quasi-periodic (see Fig.* 4.1):

$$\varphi_j(t) = \varphi_j(0) + \omega_j(I)t.$$

Example 4.3.

(a) For the van der Pol system with $\varepsilon = 0$ and $H = \frac{1}{2}(x^2 + y^2)$, the action–angle variables are: $I = H$ and $\varphi = \arg(x + iy)$.

(b) For two independent harmonic oscillators we have $H = \frac{1}{2}\omega_1\left(q_1^2 + p_1^2\right) + \frac{1}{2}\omega_2\left(q_2^2 + p_2^2\right)$. Here the functions $I_1 = \frac{1}{2}\left(q_1^2 + p_1^2\right)$ and $I_2 = \frac{1}{2}\left(q_2^2 + p_2^2\right)$ are first integrals in involution. Together with $\varphi_1 = \arg\left(q_1 + ip_1\right)$ and $\varphi_2 = \arg\left(q_2 + ip_2\right)$ they form the action–angle coordinates. The invariant tori are products of two circles $\left\{q_1^2 + p_1^2 = r_1^2\right\} \times \left\{q_2^2 + p_2^2 = r_2^2\right\}$ with the phase curves like in Fig. 4.1.

The projections of these tori to the configuration space of q's are rectangles with Lisajous curves as projections of the phase curves.

Example 4.4 (Kepler problem). In   this   problem,   the Hamiltonian is as follows:

$$H = \frac{1}{2}\left|p\right|^2 - \frac{k}{|q|}, \tag{4.3}$$

where $q = (q_1, q_2)$ are coordinates and $p = (p_1, p_2)$ are momenta and $k = GM$ is a constant (G — the gravity constant and M — the mass of heavy body at $q = 0$). In fact, since we have chosen the mass of orbiting body equal 1, the momenta p_j are the same as the velocities $\dot{q}_j$.

It is convenient to use the polar coordinates $r = |q|$, $\theta = \arg(q_1 + iq_2)$, in which the corresponding Newtonian system

$$\ddot{q} = -kq/|q|^3$$

becomes simplified. Let $e_r = q/|q| = (\cos\theta, \sin\theta)$ and $e_\theta = (-\sin\theta, \cos\theta)$ be the orthogonal unit vectors along the radii and circles. We express our vectors in the frame (e_r, e_θ). Note the formulas (Exercise 4.27)

$$\dot{e}_r = \dot{\theta}e_\theta, \quad \dot{e}_\theta = -\dot{\theta}e_r. \tag{4.4}$$

We have $q = re_r$ and hence

$$\dot{q} = \dot{r}e_r + r\dot{\theta}e_\theta,$$
$$\ddot{q} = (\ddot{r} - r\dot{\theta}^2)e_r + (2\dot{r}\dot{\theta} + r\ddot{\theta})e_\theta.$$

So, the Lagrangian equals

$$L = T - U = \frac{1}{2}\dot{r}^2 + \frac{1}{2}r^2\dot{\theta}^2 + \frac{k}{r},$$

the generalized momenta are $p_r = \partial L/\partial \dot{r} = \dot{r}$ and $p_\theta = \partial L/\partial \dot{\theta} = r^2\dot{\theta}$, and the Hamiltonian becomes

$$H = \frac{1}{2}p_r^2 + \frac{p_\theta^2}{2r^2} - \frac{k}{r}.$$

The above Newtonian system changes to

$$\frac{\mathrm{d}}{\mathrm{d}t}(r^2\dot{\theta}) = 0, \quad \ddot{r} = -\frac{k}{r} + r\dot{\theta}^2.$$

The first of them means that the generalized momentum $p_\theta = r^2\dot{\theta}$ is a constant of motion; it is known as the *angular momentum* (note that θ is a *cyclic coordinate*, see Section 6.5.1 in Appendix). Since H and p_θ are independent first integrals, the Hamiltonian system is completely integrable.

Next, assuming the value of the angular momentum $M = p_\theta \neq 0$, we have $\dot{\theta} = M/r^2$ and the second of the above equations takes the Newtonian form $\ddot{r} = -V'(r)$, where

$$V = -k/r + M^2/2r^2$$

is the *effective potential*. We have

$$V(r) \to +\infty \text{ as } r \to 0^+, \quad V(r) \to 0^- \quad \text{as } r \to +\infty$$

and $V(r)$ has unique minimum $E_{\min} = -k^2/2M^2 < 0$ at $r_{\min} = M^2/k$.

When the total energy E lies between $E_{\min}$ and 0, then the curve

$$\Gamma = \Gamma_{M,E} = \left\{ \frac{1}{2}p_r^2 + V(r) = E \right\}$$

in the (r, p_r)-plane is an oval (diffeomorphic to a circle) defined by $p_r = \pm\sqrt{2(E - V(r))}$, $r_1 < r < r_2$.

From this we can identify the common level surfaces $\{H = E, p_\theta = M\}$ as 2-tori. Indeed, one angle coordinate should be θ; note that $\dot{\theta} = M/r^2$ has constant sign. The other angle coordinate parametrizes the closed curve $\Gamma = \Gamma_{M,E}$ as the "time" (which satisfies $dt = dr/p_r$):

$$\psi = \psi(r, p_r) = \int_\gamma \frac{dr}{p_r},$$

where the integration path $\gamma = \gamma(r, p_r)$ lies in the curve Γ, starts at $(r_1, 0)$ and ends at (r, p_r). We have two *periods*:

$$T_1 = \oint_\Gamma \frac{dr}{p_r}, \quad T_2 = M \int_0^{T_1} \frac{dt}{r^2} = M \oint_\Gamma \frac{dr}{r^2 p_r}.$$

If T_1/T_2 is rational, then the evolution is periodic, otherwise it is quasi-periodic.

The angle part of the *action–angle coordinates* is as follows:

$$\varphi_1 = \frac{2\pi}{T_1}\psi, \quad \varphi_2 = \frac{2\pi}{T_2}\left(\theta - M \int_\gamma \frac{dr}{r^2 p_r} \right) + \frac{T_1}{T_2}\varphi_1. \qquad (4.5)$$

Then we get

$$\dot{\varphi}_1 = \frac{2\pi}{T_1}, \quad \dot{\varphi}_2 = \frac{2\pi}{T_2}\left(\dot{\theta} - M \cdot \frac{1}{r^2} \right) + \frac{T_1}{T_2}\dot{\varphi}_1 = \frac{2\pi}{T_2}.$$

The action part of the action–angle variables can be constructed using the property $dH = (2\pi/T_1)\,dI_1 + (2\pi/T_2)\,dI_2$, but in a non-explicit way; we skip it.

The projection of such torus into the coordinate plane is the ring with interior circle of radius r_1 and the exterior circle of radius r_2. We also know that the projections of the phase curves into the coordinate plane are ellipses (Exercise 4.28); so, the motion on this torus is periodic.

We note that the same functions, the total energy and the angular momentum, are first integrals also in the case when we replace the Coulomb potential $-k/\,|q|$ with any function $\widetilde{U}\,(|q|)$ depending only on $|q|$, i.e., a central potential. For example, we can take $\widetilde{U}\,(|q|) = -\frac{1}{2}\,|q|^2$(Exercise 4.29).

Another remark concerns the case of central potential in $\mathbb{R}^3$. Here we have the vector $M = \dot{q} \times q$ as the angular momentum. But here the Poisson brackets of the components M_i of M are not trivial.

Example 4.5 (Rigid body around a fixed point). This is the one of fundamental mechanical systems. We have a rigid body in constant gravitational field which moves around fixed point, like a top (see Fig. 4.2).

Recall that, if on a material point of mass $\mu^{(k)}$ at $x^{(k)} \in \mathbb{R}^3$ acts the gravitational force $f^{(k)} = \mu^{(k)}g\gamma$, the Newtonian equations can be rewritten in the following form:

$$\frac{d}{dt}(x^{(k)} \times p^{(k)}) = x^{(k)} \times f^{(k)}, \qquad (4.6)$$

where $\gamma = (0, 0, -1)$ is the unit vector in the direction of the gravity, g is the gravitational acceleration, $p^{(k)} = \mu^{(k)}\dot{x}^{(k)}$ is the momentum of the point and $x^{(k)} \times p^{(k)} = m^{(k)}$ is the angular momentum of the points and $x^{(k)} \times f^{(k)}$ is the moment of the gravitational force

Fig. 4.2.   Rigid body.

with respect to the point $x = 0$. In the case of a rigid body (where the distances between its points are constant) consisting of discrete points of masses $\mu^{(k)}$ at points $x^{(k)}$, we sum up Eq. (4.6) over k's and get

$$\dot{m} = x_c \times f, \qquad (4.7)$$

where $m = \sum m^{(k)}$ is the angular momentum of the body, $f = \mu g \gamma = \left(\sum \mu^{(k)} \right) g \gamma$ is the gravity force acted on thee body, $x_c = \frac{1}{\mu} \sum \mu^{(k)} x^{(k)}$ is the mass center of the body and μ is the body mass.

Recall the formula

$$\dot{x}^{(k)} = \omega \times x^{(k)}$$

for the velocity of the point $x^{(k)}$ moving around an axis $\ell = \mathbb{R}\omega \subset \mathbb{R}^3$ with the angular velocity $|\omega|$. In this case, $m^{(k)} = \mu^{(k)} x^{(k)} \times \left(\omega \times x^{(k)} \right) = I^{(k)} \omega$, where $I^{(k)}$ is the moment of inertia matrix (tensor). In the case of a continuous body, the sums are replaced with suitable integrals. For the whole body we have

$$m = I\omega,$$

where I is the inertia momentum matrix (inertia tensor of the body). Therefore, Eq. (4.7) take the form

$$\frac{\mathrm{d}}{\mathrm{d}t} (I\omega) = -\gamma \times k, \quad k = \mu g x_c.$$

Unfortunately, this system is not easy, because the quantities I, ω, k evolve, only the vector γ is constant.

To cure this deficiency one rewrites the latter system in the coordinate frame associated with the moving body. In that frame, the vectors are denoted by capital letters; we have

$$x = AX, \quad p = AP, \quad m = AM, \quad \omega = A\Omega,$$

where $A = A(t)$ is the matrix of corresponding change of coordinates. This matrix is orthogonal, $AA^\top = I$, which implies formulas of the following form:

$$\dot{x} = \dot{A}X + A\dot{X} = \dot{A}A^{-1}x + AV.$$

Here the matrix $\dot{A}A^{-1}$ is antisymmetric, because $\dot{A}A^\top + A\dot{A}^\top = 0$, and we have

$$\dot{A}A^{-1}x = \omega \times x,$$

where ω is the angular velocity of the body with respect to the constant coordinate frame. Thus, we have $\dot{x} = \omega \times x + A\dot{X}$ and hence $\dot{X} = X \times \Omega + A^{-1}\dot{x}$. Analogously, we get $\dot{M} = M \times \Omega + A^{-1}\dot{m}$, where $A^{-1}\dot{m} = -A^{-1}(\gamma \times k) = -\Gamma \times K$. Together with $\dot{\gamma} = A(\dot{\Gamma} - \Gamma \times \Omega) = 0$, we get the following **Euler–Poisson system**:

$$\dot{M} = M \times \Omega - \Gamma \times K, \quad \dot{\Gamma} = \Gamma \times \Omega. \qquad (4.8)$$

Recall that $M = I\Omega$, where I is the inertia matrix, which is symmetric and constant. We can assume that the matrix I is diagonal,

$$I = \mathrm{diag}\,(I_1, I_2, I_3).$$

Also the vector $K = \mu g X_c$ is constant and proportional to the position of X_c of the mass center of the body. Thus, we have a closed six-dimensional system for Ω, Γ, or for M, Γ.

In the space of (M, Γ)'s, we introduce the following brackets:

$$\{M_1, M_2\} = M_3, \quad \{M_2, M_3\} = M_1, \quad \{M_3, M_1\} = M_2,$$
$$\{M_1, \Gamma_2\} = \Gamma_3, \quad \{M_2, \Gamma_3\} = \Gamma_1, \quad \{M_3, \Gamma_1\} = \Gamma_2, \qquad (4.9)$$
$$\{M_i, \Gamma_i\} = 0, \quad \{\Gamma_i, \Gamma_j\} = 0.$$

Recall that the bracket is antisymmetric and extends to all polynomial functions by the Leibnitz rule $\{fg, h\} = f\{g, h\} + g\{f, h\}$. It is a Poisson bracket if its obeys the Jacobi identity(see Section 6.5.5) $\{\{f, g\}, h\} + \{\{g, h\}, f\} + \{\{h, f\}, g\} = 0$. It turns out that Eq. (4.9) define a genuine Poisson bracket. Recall also that a function h defines the Hamiltonian vector field X_h by the formula $X_h(f) = \{f, h\}$.

It turns out that the Euler–Poisson system (4.8) is Hamiltonian with respect to the Poisson structure (4.9) with the total energy as the Hamilton function

$$H = T + U = \frac{1}{2}(M, \Omega) - (K, \Gamma), \qquad (4.10)$$

where T and U denote the kinetic and potential energy. For example, $\dot{M}_i = \{M_i, H\}$ (Exercise 4.30).

There is a delicate point in this approach. The above Poisson structure is degenerate in the sense that some functions have zero Poisson brackets with all other functions. These, the so-called Casimir function, are generated by the following functions:

$$\Gamma^2 = (\Gamma, \Gamma), \quad (M, \Gamma), \tag{4.11}$$

i.e., the so-called geometrical integral and the minus the vertical component of the angular momentum. Of course, they are the first integrals for the Hamiltonian vector field X_H.

But the manifolds

$$\mathcal{Y} = \mathcal{Y}_{m_3} = \{(M, \Gamma) : \Gamma^2 = 1, \ (M, \Gamma) = -m_3\} \tag{4.12}$$

are such that the above Poisson bracket restricted to functions on $\mathcal{Y}$ is non-degenerate; these manifolds are called the symplectic leaves. This means that the vector field X_H restricted to $\mathcal{Y}$ is a genuine Hamiltonian vector field defined via a symplectic structure.

The leaves $\mathcal{Y}$ are four-dimensional and support Hamiltonian systems with two degrees of freedom. Therefore, the existence of an additional first integral, i.e., independent of Γ^2, (M, Γ) and H, ensures that the vector field $X_H|_{\mathcal{Y}}$ is completely integrable; we say that the Euler–Poisson system is completely integrable.

Integrable cases of the Euler–Poisson system were found centuries ago. We have:

- $K = 0$ (Euler);
- $I_1 = I_2$, $K_1 = K_2 = 0$ (Lagrange);
- $I_1 = I_2 - 2I_3$, $K_3 = 0$ (Kovalevskaya).

In the Euler case, the additional first integral is M^2; in the Lagrange case, we have M_3, the component of M along the symmetry axis; in the Kovalevskaya case, it is $|F|^2$, where $F = (M_1 + iM_2)^2 - 2|K| I_3 (\Gamma_1 + i\Gamma_2)$. It turns out that these are the only cases with a meromorphic first integral and this important fact was proved in the beginning of the 1980s by S. Ziglin.

The Euler case is well investigated (see [11]). Let us analyze the Lagrange case.

Recall that $K_1 = K_2 = 0$ and $I_1 = I_2$; assume $K_3 \neq 0$ (we will denote it by k) and $0 \neq I_3 \neq I_2 \neq 0$. Introduce the following notations:

$$m = M_1 + \mathrm{i}M_2 = \sqrt{x}\mathrm{e}^{\mathrm{i}\varphi}, \quad x = |m|^2 = M_1^2 + M_2^2, \quad y = (M, K \times \Gamma).$$
$$(4.13)$$

We have a family of invariant surfaces

$$\mathcal{S} = \mathcal{S}_{E,M_3} = \{(M,\Gamma) \in \mathcal{Y}_{m_3} : H = E, \ (M,K) = kM_3\}, \quad (4.14)$$

where M_3 is the component of the angular momentum along the symmetry axis.

From Eq. (4.8), we get

$$\dot{x} = 2y.$$

The kinetic energy equals $T = x/2I_2 + M_3^2/2I_3$ and the potential energy equals $U = -(\Gamma, K) = E - T$; thus $(\Gamma, K) = x/2I_2 - F$, where $F = E - M_3^2/2I_3$ is a constant.

Let us expand the vector Γ in the frame $\{M, K, M \times K\}$ (which is not orthogonal), i.e.,

$$\Gamma = \alpha M + \beta K + \gamma M \times K.$$

We have

$$-m_3 = (\Gamma, M) = \alpha\left(x^2 + M_3^2\right) + \beta k M_3,$$
$$\frac{x}{2I_2} - F = (\Gamma, K) = \alpha k M_3 + \beta k^2,$$
$$y = (\Gamma, M \times K) = \gamma\left(M \times K\right)^2 = \gamma(M^2 K^2 - (M, K)^2)$$
$$= \gamma k^2 x.$$

From this we compute the coefficients:

$$\alpha = -\frac{m_3 k + M_3\left(x/2I_2 - F\right)}{kx},$$
$$\beta = \frac{\left(x + M_3^2\right)\left(x/2I_2 - F\right) + k m_3 M_3}{k^2 x},$$
$$\gamma = \frac{y}{k^2 x}.$$

Now, the property $\Gamma^2 = 1$, i.e., $\alpha^2 \left(x + M_3^2\right) + 2\alpha\beta k M_3 + \beta^2 k\,[2] + \gamma^2 k^2 x = 1$, leads to the elliptic curve equation

$$y^2 = R(x), \qquad\qquad (4.15)$$

$R = -\left(x + M_3^2\right)\left(x/2I_2 - F\right)^2 + k^2 x - (m_3 k)^2 - 2m_3 M_3 k\left(x/2I_2 - F\right)$.

Recall that $x = M_1^2 + M_2^2 \geq 0$. We have $(x/2I_2 - F)|_{x=0} = M_3^2/2I_3 - E$ and hence $R(0) = -\left[M_3\left(M_3^2/2I_3 - E\right) + m_3 k\right]^2 \leq 0$. Since $R(X) \sim -x^3/4I_2^2$ for large x, we conclude that the only possible component of the curve $y^2 = R(x)$ is the oval

$$\Gamma = \Gamma_{M_3}$$

diffeomorphic with a circle (in a domain where $R(x) > 0$ and $x > 0$). This oval can degenerate to a point or to a separatrix loop. Therefore, assuming the oval Γ is non-degenerate, the map

$$(x, y, \varphi) : \mathcal{S} \longmapsto \mathbb{R}^2 \times \mathbb{S}^1$$

realizes a diffeomorphism between $\mathcal{S}$ and a 2-torus in its image.

Let us describe the dynamics on the invariant tori. Substituting the expanded vector Γ into the first of Eq. (4.8) we get $\dot{M} = M \times I^{-1}M - \alpha k(M_2, -M_1, 0) + \gamma k^2\left(M_1, M_2, 0\right)$. With complex $m = M_1 + iM_2$ given in Eq. (4.13) it implies

$$\dot{m} = (C + iD)\,m,$$
$$C = C(x, y) = \gamma k^2, \quad D = D(x) = \alpha k - L M_3,$$

where $\gamma = \gamma(x, y)$ depends on X, Y, $\alpha = \alpha(x)$ depends on X and $L = 1/I_3 - 1/I_2$. For the argument φ of m (in Eq. (4.13)) we get the equation

$$\dot{\varphi} = D(x) = -L M_3 - (m_3 k + M_3\left(x/2I_2 - F\right))/x.$$

Therefore, we arrive at the following **Lagrange system**:

$$\dot{x} = 2y, \quad y^2 = R(x), \quad \dot{\varphi} = D(x). \qquad (4.16)$$

The first two of Eq. (4.16) are solved by means of elliptic functions. We have

$$\int_{x_0}^{x} \frac{\mathrm{d}s}{\sqrt{R(s)}} = t;$$

thus $x = \mathcal{P}(t)$, where the so-called *Weierstrass $\mathcal{P}$-function* is the inverse of the left-hand side of the above equation. It is periodic with the *period*

$$T_1 = \frac{1}{2} \oint_\Gamma \frac{\mathrm{d}x}{y} > 0.$$

The third of Eq. (4.16) is integrated as follows:

$$\varphi(t) = \varphi_* + \int_0^t D\left(\mathcal{P}(s)\right) \mathrm{d}s$$

and important is the *monodromy shift* (or the *second period*)

$$T_2 = \int_0^{T_1} D\left(\mathcal{P}(s)\right) \mathrm{d}s = \frac{1}{2} \oint_\Gamma D(x) \frac{\mathrm{d}x}{y}.$$

Here the movement of φ is called the *precession* and the movement of x is called the *nutation*.

Depending on whether the ratio

$$\Omega = T_2/T_1$$

is rational or not the motion on the invariant surface $\mathcal{S}$ is periodic or quasi-periodic (like in the Liouville–Arnold theorem).

One can now construct the *angle–action variables*. Let $x_0 < 0 < x_1 < x_2$ be the zeroes of the cubic polynomial $R(x)$ (such that $\Gamma \subset \{x_1 < x < x_2\}$). The angles are defined as follows (cf. Eq. (4.5)):

$$\varphi_1 = \frac{\pi}{T_1} \int \frac{\mathrm{d}x}{y}, \quad \varphi_2 = \frac{\pi}{T_2} \left(\varphi - \frac{1}{2} \int D(x) \frac{\mathrm{d}x}{y} \right) + \frac{T_1}{T_2} \varphi_1,$$

where the integration path in the both integrals lies in the oval Γ, starts at $(x_1, 0)$ and ends at (x, y) (it may run several times along the whole oval). Because $\mathrm{d}t = \mathrm{d}x/2y$ on Γ, we have

$$\dot{\varphi}_1 = 2\pi/T_1, \quad \dot{\varphi}_2 = (\pi/T_2)(\dot{\varphi} - D(x)) + (T_2/T_1)\dot{\varphi}_1 = 2\pi/T_2,$$

where $T_{1,2} = T_{1,2}(H, M_3)$ depend on the values of the first integrals H and $M_3 = (M, K)$.

We also have to find the corresponding actions I_1 and I_2. We will do this in a non-explicit way. We have the condition

$$\mathrm{d}H = (2\pi/T_1)\,\mathrm{d}I_1 + (2\pi/T_2)\,\mathrm{d}I_2, \qquad (4.17)$$

i.e., $\dot{\varphi}_j = \partial H/\partial I_j$. Assume

$$\mathrm{d}I_1 = A\mathrm{d}H + B\mathrm{d}M_3, \quad \mathrm{d}I_2 = C\mathrm{d}\tilde{I}_1 + D\mathrm{d}\tilde{I}_2,$$

where $A = A(H, M_3)$, etc. Substituting this to Eq. (4.17), we find $C = T_2/2\pi - \Omega A$ and $D = -\Omega B$, $\Omega = T_2/T_1$, which reduces the problem to finding the functions $A(H, M_3)$ and $B(H, M_3)$. The closeness conditions of the 1-forms $\mathrm{d}I_j$ allows to express B in terms of A, $B = \left(\Omega'_{M_3}/\Omega'_H\right) A - \frac{1}{2\pi}(T_2)'_{M_3}/\Omega'_H$, and to arrive at the first-order PDE for A,

$$\frac{\Omega'_{M_3}}{\Omega'_H} A'_H - A'_{M_3} + \left(\frac{\Omega'_{M_3}}{\Omega'_H}\right)'_H A - \frac{1}{2\pi}\left(\frac{(T_2)'_{M_3}}{\Omega'_H}\right)'_H = 0,$$

which has a solution (at least locally at a generic point).

More known approach to the Lagrange case uses the parametrization of $SO(3)$, the configuration space of the body, by the Euler angles φ, ψ, θ (see [11]). One has the Lagrange function

$$L = \frac{I_2}{2}(\dot{\theta}^2 + \dot{\varphi}^2 \sin^2 \theta) + \frac{I_3}{2}(\dot{\psi} + \dot{\varphi}\cos\theta)^2 - k\cos\theta.$$

Since φ and ψ are cyclic coordinates, the corresponding (angular) momenta

$$p_\varphi = \frac{\partial L}{\partial\dot{\varphi}} = \dot{\varphi}\left(I_2 \sin^2\theta + I_3 \cos^2\theta\right) + \dot{\psi}I_3 \cos\theta = m_3,$$

$$p_\psi = \frac{\partial L}{\partial\dot{\psi}} = \dot{\varphi}I_3 \cos\theta + \dot{\psi}I_3 = M_3$$

are first integrals. Calculating from this the $\dot{\varphi}$ and $\dot{\psi}$ one obtains the total energy

$$H = \frac{I_2}{2}\dot{\theta}^2 + \frac{(m_3 - M_3 \cos\theta)^2}{2I_2 \sin^2\theta} + \frac{M_3^2}{2I_3} + k\cos\theta.$$

and

$$\dot{\varphi} = \frac{m_3 - M_3 \cos\theta}{I_2 \sin^2\theta}, \quad \dot{\psi} = \frac{I_2 M_3 \sin^2\theta + I_3 \cos\theta \, (M_3 \cos\theta - m_3)}{I_2 I_3 \sin^2\theta}.$$

With variables $v = \cos\theta$ and $w = \dot{v}$ the corresponding Lagrange system takes the form

$$\dot{v} = w, \qquad w^2 = (c - dv)\left(1 - v^2\right) - (e - fv)^2,$$

$$\dot{\varphi} = \frac{e - fv}{1 - v^2}, \qquad \dot{\psi} = \frac{g - hv + lv^2}{1 - v^2}$$

with some constants $c, \ldots, l$. The elliptic curve in the second of the above equations contains a compact oval Γ in $\{-1 \le v \le 1\}$ and the map (v, w, φ, ψ) from $\{p_\varphi = m_3, p_\psi = M_3, H = E\}$ to $\mathbb{R}^2 \times \mathbb{S}^2 \times \mathbb{S}^1$ has a 3-torus as its image.

But here we have three degrees of freedom and the levels of the Hamilton function $\{H = E\}$ are five-dimensional. Therefore, the 3-tori from the Liouville–Arnold theorem do not divide the latter hypersurfaces into invariant domains. Therefore, the approach via the degenerate Poisson structure and the Euler–Poisson equations is more useful.

4.1.2. *Averaging theorem*

Consider now the following perturbation of system (4.2):

$$\dot{I} = \varepsilon g(I, \varphi), \quad \dot{\varphi} = \omega(I) + \varepsilon f(I, \varphi), \tag{4.18}$$

where $I = (I_1, \ldots, I_n)$, $\varphi = (\varphi_1, \ldots, \varphi_n)$ and $\omega = (\omega_1, \ldots, \omega_n)$. In fact, we do not need to assume that the number of I_j's is the same as the number of φ_j's, so we will assume $I = (I_1, \ldots, I_m) \in \mathbb{R}^m$.

It is natural to expect that the solution of system (4.18) after time of order $O(1)$ diverges from the solution of system (4.2) with the same initial conditions at a distance of order $O(\varepsilon)$. On the other hand, the following theorem says that the same distance $O(\varepsilon)$ can be obtained after a time that tends to infinity as $\varepsilon \to 0$. This kind of phenomenon takes place due to the so-called averaging.

The idea of averaging is related with the fact that on typical torus $\mathbb{T}^n = \{I = d\}$ the trajectories of the non-perturbed system are dense (like in Fig. 4.1). Therefore, the average deviation of the action $I(t)$ can be calculated approximately by averaging on the torus.

We define the **averaged system**

$$\dot{J} = \varepsilon G(J), \qquad (4.19)$$

where

$$G(J) = \left(\frac{1}{2\pi}\right)^n \int_0^{2\pi} \dots \int_0^{2\pi} g(J, \varphi) \mathrm{d}\varphi_1 \dots \mathrm{d}\varphi_n$$

is the averaged over $\mathbb{T}^n$ speed of change of the action.

Theorem 4.6 (About averaging). *Let $n = 1$ and the functions ω, f, g be of class C^1 and $\omega(I) > 0$ on an open subset $U \times \mathbb{T}^1 \subset \mathbb{R}^1 \times \mathbb{T}^1$. If $(I(t), \varphi(t))$ and $(J(t), \psi(t))$ are solutions of the systems (4.18) and (4.19) such that $I(0) = J(0)$ then, for*

$$0 < t < 1/\varepsilon,$$

we have

$$|I(t) - J(t)| < C \cdot \varepsilon,$$

where constant C depends only on ω, f, g.

Proof. Let us apply such change

$$K = I + \varepsilon k(I, \varphi) \qquad (4.20)$$

that one would get $\dot{K} = O(\varepsilon^2)$. The calculation of $k(J, \varphi)$ runs as follows:

$$\dot{K} = \dot{I} + \varepsilon \frac{\partial k}{\partial I} \dot{I} + \varepsilon \frac{\partial k}{\partial \varphi} \dot{\varphi} = \varepsilon \left\{ g + \frac{\partial k}{\partial \varphi} \omega \right\} + O(\varepsilon^2)$$
$$= \varepsilon \left\{ g(K, \varphi) + \frac{\partial k}{\partial \varphi}(K, \varphi) \omega(K) \right\} + O(\varepsilon^2).$$

So, we want to solve the equation

$$\frac{\partial k}{\partial \varphi}(K, \varphi) \omega(K) = -g(K, \varphi)$$

with the obvious solution $k(K, \varphi) = \frac{-1}{\omega(K)} \int_0^{\varphi} g(K, \psi) \mathrm{d}\psi$. Unfortunately, usually this solution is not a single valued (not periodic) function of the variable φ. The obstacle is the quantity $\int_0^{2\pi} g(h, \psi) d\psi$, which can be non-zero.

But, by writing

$$g(K, \varphi) = G(K) + \tilde{g}(K, \varphi),$$

such that $\int_0^{2\pi} \tilde{g}(h, \psi) d\psi = 0$, we can define the single-valued function

$$k(K, \varphi) = \frac{-1}{\omega(K)} \int_0^\varphi \tilde{g}(K, \psi) d\psi.$$

We get the equation

$$\dot{K} = \varepsilon G(K) + O(\varepsilon^2).$$

You can see that, after the time $O(1/\varepsilon)$, the difference between $J(t)$ and $K(t)$ is of order $O(\varepsilon)$. On the other hand, the difference between $K(t)$ and $I(t)$ is of order $O(\varepsilon)$, due to change (4.20). $\square$

For perturbations of type (4.18) of completely integral Hamiltonian systems with many degrees of freedom the estimates are weaker than in the thesis of Theorem 4.6. It turns out that, after time of order $O(1/\varepsilon^a)$ and for initial conditions outside a set with Lebesque measure $O(\varepsilon^b)$, the divergence of $J(t)$ from $I(t)$ does not exceed $O(\varepsilon^c)$, where $a, b, c > 0$ are exponents depending on ω, f, g. For more information, we refer the reader to [6].

The averaging like in Theorem 4.6 was used in studying some codimension two bifurcations (see Secs. 3.8.2 and 3.8.4 in the previous chapter). In Section 3.8.3, we used averaging over a two-dimensional torus.

4.1.3. *Abelian integrals*

Important is the following application of the averaging in the problem of limit cycles for planar vector fields. Let us consider again a perturbation of a two-dimensional Hamiltonian system:

$$\dot{x} = H_y' + \varepsilon P(x, y), \quad \dot{y} = -H_x' + \varepsilon Q(x, y). \tag{4.21}$$

For $\varepsilon = 0$ the phase curves lie in the levels of Hamilton function $H(x, y)$. In some region of the phase space, these curves are closed.

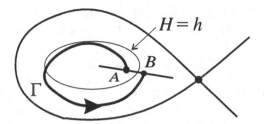

Fig. 4.3. Return map for perturbation of Hamiltonian system.

As we have already done several times, the study of limit cycles of the perturbed system relies on analyzing the Poincaré return map from S (transversal section to phase curves) to S. Parametrizing S with $H|_S$ the limit cycle condition is $\Delta H = H(B) - H(A) = 0$ (see Fig. 4.3). We have

$$
\begin{aligned}
\Delta H &= \int_0^T \frac{\mathrm{d}H}{\mathrm{d}t}\mathrm{d}t = \int_0^T \left\{ H'_x \left(H'_y + \varepsilon P \right) + H'_y (-H'_x + \varepsilon Q) \right\} \mathrm{d}t \\
&= \varepsilon \int \left(PH'x + QH'_y \right) \mathrm{d}t = \varepsilon \int \left\{ P(H'_x - \varepsilon Q) + Q \left(H'_y + \varepsilon P \right) \right\} \mathrm{d}t \\
&= \varepsilon \int_{\Gamma(h)} Q\mathrm{d}x - P\mathrm{d}y = \varepsilon \oint_{H=h} (Q\mathrm{d}x - P\mathrm{d}y) + O(\varepsilon^2),
\end{aligned}
$$

where T is the return time to S and $\Gamma(h)$ is the phase curve of the perturbed system starting from $A \in S$, such that $H(A) = h$, and ending at $B \in S$.

(There exists a simpler derivation of this formula. Note that the phase curves of the vector field (4.21) satisfy the following Pfaffian equation:

$$
\left(H'_y + \varepsilon P \right) \mathrm{d}y + \left(H'_x - \varepsilon Q \right) \mathrm{d}x = \mathrm{d}H - \varepsilon \left(Q\mathrm{d}x - P\mathrm{d}y \right) = 0,
$$

i.e., the 1-form on the left-hand side vanishes after restricting to a phase curve.)

The expression

$$
I(h) = \oint_{H=h} Q\mathrm{d}x - P\mathrm{d}y \tag{4.22}
$$

is the so-called **Abelian integral**[a] (or the *Melnikov function*). It is not difficult to see that the function $I(h)$ is equivalent to the function $G(J)$ from Eq. (4.19).

This leads to the following proposition.

Proposition 4.7. *If $I(h_0) = 0$ and $I'(h_0) \neq 0$, then for small $\varepsilon \neq 0$ there exists a limit cycle γ_ε which tends to the curve $H = h_0$ as $\varepsilon \to 0$.*

Proof. Use the Implicit Function Theorem for the equation $\Delta H / \varepsilon = 0$. $\qquad\square$

Remark 4.8.

(a) Proposition 4.7 can be generalized to the case when the system (4.21) and the integral (4.22) depend on some additional parameters. For example, if it depends on μ, then we allow the situation with $I(h_0; \mu_0) = I'_h(h_0; \mu_0) = 0$ but $I''_{hh}(h_0; \mu_0) \neq 0$; then two limit cycles $\gamma'_\varepsilon(\mu)$ and $\gamma''_\varepsilon(\mu)$ collide at a semi-stable limit cycle close to $\{H = h_0\}$ for μ close to μ_0.

(b) Sometimes we deal with perturbations of integrable systems which are not Hamiltonian, but are of the form

$$\dot{x} = M^{-1}H'_y + \varepsilon P, \quad \dot{y} = -M^{-1}H'_x + \varepsilon Q, \qquad (4.23)$$

i.e., with the integrating factor M for $\varepsilon = 0$. Then the integral (4.22) is replaced with the integral

$$I(h) = \oint M(Q\mathrm{d}x - P\mathrm{d}y). \qquad (4.24)$$

The approach to the problem of limit cycles using Abelian integrals is common in the Qualitative Theory. Recall that in the weakened 16th Hilbert problem one asks about an estimate of the number of zeros of the number of zeros of the Abelian integral in terms of $n = \max(\deg H, \deg P, \deg Q)$. Recall also that there is a

---

[a]The concept of Abelian integrals comes from the complex algebraic geometry. They are integrals of meromorphic 1-forms along certain closed curves on complex algebraic curves (Riemann surfaces). When $H(x,y)$, $P(x,y)$ and $Q(x,y)$ are polynomials, then the Riemann surface is the complex curve $\{H(x,y) = h\} \subset \mathbb{C}^2$ and the 1-form is $\omega = (Q\mathrm{d}x - P\mathrm{d}y)|_{H=h}$.

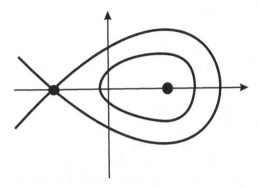

Fig. 4.4.   Elliptic Hamiltonian.

double exponential estimate for this number (see Remark (2.19) in Chapter 2). But there are no general estimates in the case of integrals (4.24).

Even for fixed Hamiltonian the problem is not easy. Of course, for quadratic H the integral is computed explicitly (Exercise 4.32). In the following, we present estimates for simplest cubic Hamiltonian.

Example 4.9 (Elliptic integrals). The Hamiltonian takes the form

$$H = y^2 + x^3 - x \qquad (4.25)$$

and its level curves are elliptic curves (Fig. 4.4).

We deal with the integrals

$$I_\omega(h) = \oint_{\gamma_h} \omega,$$

where $\omega = Q\mathrm{d}x - P\mathrm{d}y$ is a polynomial 1-form of degree $n = \max(\deg P, \deg Q)$ and γ_h is a compact component of the real elliptic curve $\{H(x, y) = h\}$. Since the critical points of H are $p_{1,2} = (\pm 1/\sqrt{3}, 0)$ with the critical values $h_{1,2} = \mp 2/3\sqrt{3}$, the domain of definition of the function $I_\omega(h)$ is the interval $[h_1, h_2]$.

Define the integrals

$$I_j = \oint_{\gamma_h} x^j y \mathrm{d}x.$$

Note the following relations on the curve γ_h:

$$x^k y^l dy = \frac{1}{l+1} d\left(x^k y^{l+1}\right) - \frac{k}{l+1} x^{k-1} y^{l+1} dx, \qquad (4.26)$$

$$y^2 = x - x^3 + h,$$

$$3x^2 dx = dx - 2y dy. \qquad (4.27)$$

The first of them allows to reduce the situation to the case with $\omega = R(x,y)\, dx$, $\deg R \leq n$. Next, the case with even powers of y in $R(x,y)$ gives zero integral. Thus, we are left with the integrals

$$\oint x^j y^{2k+1} dx = \oint x^j \left(x - x^3 + h\right)^k y dx;$$

it is a combination of functions $h^l I_m(h)$ with $3l + m \leq 3k + j$.

Next, by the integration by parts formula (4.26), we have

$$I_l = \oint x^l y dx = \frac{1}{3} \oint x^{l-2} y (dx - 2y dy) = \frac{1}{3} I_{l-2} - \frac{2}{9} \oint x^{l-2} dy^3$$

$$= \frac{1}{3} I_{l-2} + \frac{2(l-2)}{9} \oint \left(x - x^3 + h\right) x^{l-3} y dx$$

$$= \left(\frac{1}{3} + \frac{2(l-2)}{9}\right) I_{l-2} - \frac{2(l-2)}{9} I_l + \frac{2(l-1)}{9} h I_{l-3}.$$

This leads to the recurrence

$$I_l = c_1(l) I_{l-2} + c_2(l) h I_{l-3}$$

and to the formula

$$I_\omega = P_0(h) I_0 + P_1(h) I_1 \qquad (4.28)$$

with polynomials P_0 (of degree $\leq \lfloor \frac{n-1}{2} \rfloor$) and P_1 (of degree $\leq \lfloor \frac{n}{2} \rfloor + 1$). The space $W_n = \{I_\omega : \deg \omega \leq n\}$ has dimension n.

Now, we will differentiate the Abelian integrals. We have

$$I'_\omega = \oint \frac{d\omega}{dH}, \qquad (4.29)$$

where $\eta = \frac{d\omega}{dH}$ is a 1-form (Gelfand–Learay form) such that $dH \wedge \eta = d\omega$ (Exercise 4.33). η is defined modulo dH, but on γ_h is defined uniquely. In the case $\omega = x^j y dx$, we get

$$I'_j = \frac{1}{2} \oint \frac{x^j \, dx}{y}.$$

In the same way as Eq. (4.28) was derived, we prove the following relations:

$$5I_0 = 6hI'_0 + 4I'_1, \quad 21I_1 = 4I'_0 + 18hI'_1, \tag{4.30}$$

which implies

$$7 \left(27h^2 - 4 \right) I''_1 = 21I_1 \tag{4.31}$$

(Exercise 4.34).

Next, we have to extend our objects (curves, forms and integrals) into complex domain. The equation $\{H(x, y) = h\}$ defines a complex algebraic curve in $\mathbb{C}^2$, it is the Riemann surface S_h of the algebraic function $y = \sqrt{x - x^3 + h}$. This function has three branching points $x_1(h)$, $x_2(h)$ and $x_3(h)$; for $h_1 < h < h_2$ we have $x_3 < x_1 < x_2$ and $x \in [x_1, x_2]$ when $(x, y) \in \gamma_h$. In the Riemann surface S_h, the curve γ_h is homotopically equivalent to a lift to S_h of a loop θ_h in $\mathbb{C} \setminus \{x_1, x_2, x_3\}$ which surround the points x_1 and x_2.

As $h \to h_1$ the points $x_1(h)$ and $x_2(h)$ tend to $x = 1/\sqrt{3}$ and as h turns around h_2 (along a small circle) the points x_1 and x_2 exchange their position, but θ_h does not change. This implies that the integrals $I_\omega(h)$ are single valued near h_1, and $I_\omega(h_1) = 0$. Moreover, I_0 and I_1 have simple zero at h_1.

But when $h \to h_2$ the points $x_2(h)$ and $x_1(h)$ tend to the critical point $x = -1/\sqrt{3}$ and the points x_3, x_1 exchange their positions as h makes a full turn around h_2. The loop θ_h changes to $\theta_h + \vartheta_h$, where ϑ_h surrounds x_3 and x_1; in the surface S_h we have $\gamma_h \longmapsto \gamma_h + \delta_h$, where δ_h is the lift of ϑ_h. But the cycle δ_h is invariant with respect to the turning of h around h_2. This implies that the integral I_ω is ramified at h_2, but the analogous integral $J_\omega(h) = \oint_{\delta_h} \omega$ is holomorphic near h_2.

As $I_\omega(h)$ is multivalued, we consider it in the domain

$$D = \mathbb{C} \setminus [h_2, \infty].$$

As h tends in D to $h_0 > h_2$ from $\text{Im}\, h > 0$ (and from $\text{Im}\, h < 0$) it acquires some imaginary part, which equals

$$\frac{i}{2} J_\omega(h_0);$$

because the difference $I_\omega\left(h_2 + i\varepsilon e^{2\pi i}\right) - I_\omega\left(h_2 + i\varepsilon\right) = J_\omega(h_2 + i\varepsilon)$, $\varepsilon > 0$.

As $h \to \infty$ (along some ray) then, on θ_h we have $x \sim h^{1/3}$, $y \sim h^{1/2}$ and hence

$$I_0(h) \sim h^{5/6}, \quad I_1(h) \sim h^{7/6}.$$

Next, we claim that $I_1(h)$ has only one zero, at h_1, in the interval $(-\infty, h_2)$. Indeed, by Eq. (4.31) the functions $I_1(h)$ and $I_1''(h)$ have the same sign for $h < h_1 = -2/3\sqrt{3}$; so, I_1 cannot vanish there. For $h_1 < h < h_2$ the integral $I_1(h) = \int \int x dy dx$ is proportional to the center of mass of the domain bounded by γ_h; from geometry of this domain (see Fig. 4.4) it follows that this center of mass is positive.

One also finds that

$$\text{Im}\, I_1(h) \neq 0, \quad h > h_2,$$

because $\text{Im}\, I_1$ satisfies the same Eq. (4.16).

One can say more:

I_1 as only one zero (at h_1) in the domain D.

For this we use the argument principle for the contour Γ in D consisting of a large circle $\{|h| = R\}$, a small circle $\{|h - h_2| = r\}$ around h_2 and two segments along ridges of the cut $[h_2, \infty]$ (see Fig. 4.5). The increment of $\arg I_1$ along the large circle is defined by the asymptotic at infinity and equals $\frac{7}{6} \times 2\pi$. As $\text{Im}\, I_1$ does not vanish along the ridges of the cut, the increment of the argument along these segments and the small circle is $\leq 2\pi$. So, $\Delta \arg_\Gamma I_1 < 3 \times 2\pi$.

Next, we claim that:

The ratio $I_0(h)/I_1(h)$ cannot take real values for $h \in (h_2, \infty)$.

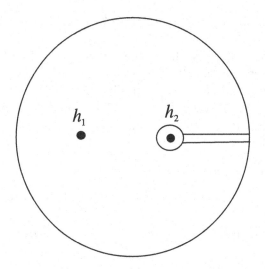

Fig. 4.5. Contour.

Indeed, if that had happened for h_0 then the vectors $(\operatorname{Re} I_0, \operatorname{Re} I_1)$ and $(\operatorname{Im} I_0, \operatorname{Im} I_1)$ would be parallel for $h = h_0$. But the both vector-valued functions form fundamental systems of solutions to the linear system (4.30). So, they should satisfy this for all $h > h_2$, i.e., $(I_0/I_1)(h) \in \mathbb{R}$ for all $h > h_2$. By the Schwarz reflection principle I_0/I_1 could be prolonged to an analytic function outside h_2, where the singularity at h_2 is removable. So, it would be entire, but it contradicts its asymptotic $\sim h^{1/3}$ at infinity.

To obtain the estimate of the number of zeroes $I_\omega(h)$ for $h_1 < h < h_2$ we estimate the number of zeroes of the function

$$I_\omega/I_1 = P_0(h)(I_0/I_1) + P_1(h)$$

in the domain D. Again, we use the argument principle with the same contour Γ. Then the increment of the argument along the large circle is $\sim \max(\deg P - 1/3, \deg P_1) \times 2\pi$. Along the ridges of the cut, the number of turns of I_ω/I_1 is bounded by the number of zeroes of the imaginary part of this function plus 1; this gives $\leq (\deg P_0 + 1) \times 2\pi$.

Summing up we get

$$\#\{h : I_\omega(h) = 0\} \leq n - 1. \tag{4.32}$$

Since the space W_n of these integrals has dimension n, a typical function from it has at least $n - 1$ zeroes. Thus, the bound (4.32) is optimal. Such spaces are called the *Chebyshev spaces* in the literature.[b]

Finally, we can apply these arguments to the Bogdanov–Takens bifurcation from Section 3.8.1, i.e., to the Abelian integral $I(h) = \oint (\nu_1 + \nu_2 x)\,y\mathrm{d}x = \nu_1 I_0 + \nu_2 I_1$, but with $H = \frac{1}{2}y^3 + \frac{1}{3}x^3 - x$ (Exercise 4.35). Anyway, we have $I_\omega/I_1 = \nu_1 I_0/I_1 + \nu_2$ and the above argument principle gives 1 as the maximal number of zeroes.

4.1.4. *Subharmonic orbits*

Here we consider non-autonomous systems of the form

$$\dot{x} = H'_y + \varepsilon P(x,y,t), \quad \dot{y} = -H'_x + \varepsilon Q(x,y,t), \tag{4.33}$$

where $H(x,y)$ is a Hamiltonian of the unperturbed system and the functions P and Q are periodic in t with period $T > 0$.

We can treat this system as a subsystem of the following four-dimensional system:

$$\dot{x} = H'_y + \varepsilon \tilde{P}(x,y,u,v), \dot{y} = -H'_y + \varepsilon \tilde{Q}(x,y,u,v),$$

$$\dot{u} = -\omega v, \dot{v} = \omega u$$

with $\omega = \frac{2\pi}{T}$. We assume that the functions P and Q are expanded into power series in $\sin \omega t$ and $\cos \omega t$. Assuming that the level curves $\{H(x,y) = h\}$ contain compact ovals γ_h, the unperturbed version of the latter system has the invariant tori $\gamma_h \times \{u^2 + v^2 = r^2\}$; in the case of Eq. (4.33) we have the tori $\mathbb{T}_h = \gamma_h \times \{u^2 + v^2 = 1\}$.

The motion on the tori $\mathbb{T}_h$ can be periodic or quasi-periodic. In this section, we are interested in the periodic orbits. Let

$$P(h) = \oint_{\gamma_h} \frac{\mathrm{d}x}{H'_y} = -\oint \frac{\mathrm{d}y}{H'_x}$$

be the periods of the solutions on the curves γ_h.

---

[b]The above proof was published by G. S. Petrov, The Chebyshev property of elliptic integrals, *Funct. Anal. Appl.* 22 (1988), 72–73.

Assume that $P(h_0) = P_0$ is commensurable with T for some h_0, i.e.,

$$P_0 = pT/q, \quad p, q \in \mathbb{N}, \quad \gcd(p, q) = 1.$$

Let $z = (x, y) = \varphi(t)$ be some solution of the unperturbed system (4.33), i.e., for $\varepsilon = 0$, corresponding to the closed phase curve γ_{h_0}, with initial condition $\varphi(0) = z_0 \in \gamma_{h_0}$. Other such solutions are

$$z = \varphi^{(s)}(t) = \varphi(t - s)$$

with varying initial condition $\varphi^{(s)}(0) = \varphi(-s) \in \gamma_{h_0}$.

We expect that, generically, for $\varepsilon \neq 0$ there should exist isolated periodic solutions of period $pT = qP_0$ located near γ_{h_0}. Such solutions are called the **subharmonic solutions**.

We have another approach to the periodic non-autonomous system (4.33). For its study, it is convenient to work with the **monodromy map** (after the period)

$$\mathcal{P} : \mathbb{R}^2 \longmapsto \mathbb{R}^2, \quad \mathcal{P} = g_0^T,$$

where $g_s^t(u_0) = \phi(t; u_0, s)$, $\phi(s; u_0, s) = u_0$, is a 2-parameter family of diffeomorphisms defining the evolution and $\mathbb{R}^2$ is identified with a section $\{s\} \times \mathbb{R}^2$ in the extended phase space $(\mathbb{R}/T\mathbb{Z}) \times \mathbb{R}^2$ (Exercise 4.36). The fixed points of this map correspond to periodic solutions of period T and the periodic points of period p correspond to periodic solutions of period pT. We look for periodic points of period p located near z_0.

Now, we can state the main result of this section. Define the following **Melnikov function**:

$$M(s) = \int_0^{pT} \left[H_x' P - H_y' Q \right] (\varphi^{(s)}(t), t) dt. \tag{4.34}$$

Theorem 4.10. *Assume that*

(i) $(\mathrm{d}P(h)/\mathrm{d}h)(h_0) \neq 0$;
(ii) *the function $M(s)$ has a simple zero s_0, $M(s_0) = 0$ and $M'(s_0) \neq 0$.*

Then there exists a family

$$z = \varphi_\varepsilon(t)$$

of solutions to system (4.33) *which are periodic of period pT and tend to $z = \varphi^{(s_0)}(t)$ as $\varepsilon \to 0$.*

Proof. Without loss of generality, we can assume that $s_0 = 0$. Consider the pth iteration of the above monodromy map, i.e.,

$$F_\varepsilon = g_0^{pT} : \mathbb{R}^2 \longmapsto \mathbb{R}^2.$$

We study this map in a neighborhood of the point $z_0 = \varphi^{(0)}(0)$. Of course, z_0 is a fixed point of the map F_0 (like other points from γ_{h_0} near z_0).

Choose a segment S transversal to the closes phase curve γ_{h_0}. For z near S we have $z = g_0^\tau(\tilde{z})|_{\varepsilon=0}$ for $\tilde{z} \in S$; $-\tau$ is the time to reach S from z along the phase curve γ_h through z. Let $\sigma = H(z) - h_0$.

The two functions $\tau = \tau(z)$ and $\sigma = \sigma(z)$ form a local coordinate system near z_0.

We express the map F_ε in these coordinates. The increment of H along the trajectory of system (4.33) equals

$$\Delta H = \Delta\sigma = \varepsilon\{M(\tau) + O(\sigma) + O(\varepsilon)\} = \varepsilon\{a\tau + o(1)\},$$

where $a = M'(0) \neq 0$. Next, it is not difficult to see that

$$\Delta\tau = -b\sigma + o(1),$$

where $b = p(\mathrm{d}P(h)/\mathrm{d}h)(h_0) \neq 0$. Therefore, the linearization matrix of the above map takes the form

$$DF_\varepsilon(z_0) = I + \begin{pmatrix} 0 & b \\ \varepsilon a & O(\varepsilon) \end{pmatrix} + h.o.t.$$

and is invertible. Now, the result follows form the Implicit Function Theorem. □

In view of the above theorem, important is the monotonicity property of the period function $P(h)$, which is also an Abelian integral.

4.2. KAM theory

Consider the Hamiltonian system

$$\dot{q} = H'_p, \quad \dot{p} = -H'_q,$$

$p = (p_1, \ldots, p_n)$, $q = (q_1, \ldots, q_n)$, with a Hamiltonian of the form

$$H(p, q) = H_0(p, q) + \varepsilon H_1(p, q),$$

where H_0 is a Hamiltonian of a completely integrable system, i.e., a system of type (4.2) in the action–angle variables. This situation is connected with one of the most important mathematical theorems of the second half of the 20th century. Before formulating it, we have to introduce two more assumptions about non-degeneration of the unperturbed Hamiltonian H_0:

$$\det\left(\frac{\partial \omega_i}{\partial I_j}\right) = \det\left(\frac{\partial^2 H_0}{\partial I_i \partial I_j}\right) \neq 0, \tag{4.35}$$

$$\det\begin{bmatrix} \partial^2 H_0/\partial I_i \partial I_j & \partial H_0/\partial I_i \\ (\partial H_0/\partial I_j)^\top & 0 \end{bmatrix} \neq 0. \tag{4.36}$$

Condition (4.35) means that the frequencies $\omega_i(I)$ vary independently and rather quickly with changes of the actions I_j, while the condition (4.36) means that these frequencies vary quickly and independently, after restriction to the levels $\{H_0 = \text{const}\}$.

Theorem 4.11 (Kolmogorov–Arnold–Moser). *If the non-degeneration conditions* (4.35) *and* (4.36) *are satisfied for H_0, then for a small perturbation $H = H_0 + \varepsilon H_1$ most of the invariant tori $\{I = \text{const}\}$ do not disappear, but become only slightly deformed, and the movement on them is still quasi-periodic.*

This theorem was formulated in 1954 at the International Congress of Mathematicians in Amsterdam, but it had to wait for a full proof until the beginning of the sixties. It was given by V. Arnold (in the analytical case) and by J. Moser (in the case of C^{333} class of smoothness). The smoothness class was later reduced to C^3.

Of course, we are not able to present this proof here; we can only say that it uses changes of variables, like in the proof of Theorem 4.6 (but with much more complicated estimates).

Example 4.12 (Planar restricted three body problem). The **planar restricted three body problem** is a system in which two bodies (interacting via the forces of gravity) rotate at a constant angular velocity around their center of mass (at the origin of the coordinate system) and the third body moves in the plane of rotation of these two bodies and has the mass so small that it does not interfere with their movement. In Fig. 4.6, we have a system in which S means Sun, J means Jupiter, and A is an Asteroid. Units of time, length and mass can be chosen so that the angular velocity, the sum of the masses of S and J and the gravitational constant are equal to 1. Then the distance between S and J also equals 1. The only parameter characterizing the system is the mass of Jupiter μ.

The equations of motion of Asteroid are Hamiltonian with the Hamilton function

$$\frac{1}{2}(p_x^2 + p_y^2) - \frac{1 - \mu}{\rho_1} - \frac{\mu}{\rho_2},$$

where ρ_1 and ρ_2 are the distances of A from S and J, respectively, and $p_{x,y}$ are momenta associated with the coordinates x, y. Note that the positions S and J change with time: $J = (1 - \mu)(\cos t, \sin t)$, $S = (-\mu)(\cos t, \sin t)$; therefore, the Hamiltonian depends directly on time.

To get rid of this dependence on time, we make the following change (simultaneous rotation of coordinates and momenta):

$$q = M(t)\,(x, y)^\top, \quad p = M(t)(p_x, p_y)^\top, \quad M(t) = \begin{pmatrix} \cos t & \sin t \\ -\sin t & \cos t \end{pmatrix}.$$

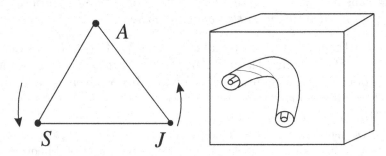

Fig. 4.6.   Three body problem and invariant tori.

It turns out that, in the new variables, the differential equations are still Hamiltonian with the new Hamiltonian

$$H = \frac{1}{2}\left(p_1 + q_2\right)^2 + \frac{1}{2}\left(p_2 - q_1\right)^2 - V(q_1, q_2), \qquad (4.37)$$

$$V = \frac{1}{2}\left(q_1^2 + q_2^2\right) + \frac{1 - \mu}{\rho_1} + \frac{\mu}{\rho_2},$$

$\rho_1^2 = (q_1 + \mu)^2 + q_2^2$, $\rho_2^2 = (q_1 + \mu - 1)^2 + q_2^2$ (Exercise 4.37). In the new variables q_1, q_2, the bodies S and J are resting.

The equilibrium points of the latter Hamiltonian system are the critical points of the Hamiltonian function (Exercise 6.61). In the case of Hamiltonian (4.37), these points, which we call the **libration points** (or the *relative equilibria*), are given by

$$p_1 = -q_2, \quad p_2 = q_1, \quad \partial V/\partial q_1 = \partial V/\partial q_2 = 0.$$

We have

$$\partial V/\partial q_2 = q_2 \left\{ 1 - (1 - \mu)\rho_1^{-3} - \mu\rho_2^{-3} \right\} = q_2 f,$$

$$\partial V/\partial q_1 = q_1 f - \mu(1 - \mu)\left(\rho_1^{-3} - \rho_2^{-3}\right).$$

We have two possibilities (Exercises 4.38 and 4.39):

(1) $q_2 = 0$; here we find three points, the so-called **collinear libration points** L_1, L_2, L_3, which are unstable.

(2) $f = 0$ and $\rho_1 = \rho_2 = 1$; here we have two so-called **triangular libration points** L_4 and L_5, which lie in the vertices of two equilateral triangles with the base $\overline{SJ}$.

Calculations show that for $27\mu(1 - \mu) > 1$ the points $L_{4,5}$ are unstable, whereas in the opposite case, i.e., for $\mu < \mu_1 = \frac{1}{2}(1 - \sqrt{23/27}) \approx 0.03852$, the eigenvalues of the linear part of the Hamiltonian system are of the form $\pm i\omega_1$, $\pm i\omega_2$, where $\omega_1 < 0 < \omega_2 \neq |\omega_1|$ (Exercise 4.40). We are on the boundary of the stability domain.

Moreover, the quadratic part of the Taylor expansion of H at point L_4 takes the form

$$H_0 = \frac{1}{2}\omega_1(\tilde{p}_1^2 + \tilde{q}_1^2) + \frac{1}{2}\omega_2\left(\tilde{p}_2^2 + \tilde{q}_2^2\right)$$

in an appropriate coordinate system in a neighborhood of L_4. It is a completely integrable Hamiltonian system with the action–angle variables $I_1 = \frac{1}{2}(\tilde{p}_1^2 + \tilde{q}_1^2)$, $I_2 = \frac{1}{2}\left(\tilde{p}_2^2 + \tilde{q}_2^2\right)$, $\varphi_1 = \arg(\tilde{q}_1 + i\tilde{p}_1)$, $\varphi_2 = \arg(\tilde{q}_2 + i\tilde{p}_2)$ and with $H_0 = \omega_1 I_1 + \omega_2 I_2$.

We have a situation like in the KAM theorem: $H = H_0 + H_1$, where H_0 is completely integrable and H_1 contains terms of order > 1 with respect to I_j's (which are small). Unfortunately, this is not enough, because the frequencies of $\omega_j = \partial H_0 / \partial I_j$ are constant, and according to the non-degeneracy condition (4.36) they should change with I_j's. Therefore, we should take into account further terms of the expansion of H in a neighborhood of L_4.

More precisely, we simplify the third- and fourth-order terms in the Hamiltonian H. This simplification is an analogue of the Poincaré–Dulac normal form and has been realized by G. Birkhoff in Theorem 4.14. This **Birkhoff normal form** in our case is as follows (Exercise 4.41):

$$H = H_0 + H_1, \quad H_0 = \omega_1 I_1 + \omega_2 I_2 + \sum \omega_{ij} I_i I_j, \quad I_j = \frac{1}{2}(P_j^2 + Q_j^2),$$

$$(4.38)$$

where $P_j = \tilde{p}_j + \cdots$, $Q_j = \tilde{q}_j + \cdots$ are new variables and H_1 contains terms of order ≥ 5 (H_0 and H_1 are different than above). In the assumption of the Birkhoff theorem, there is a condition of no resonant relations of order 3 and 4. It turns out that such relations occur for the values of $\mu_2 = \frac{1}{2}\left(1 - \sqrt{1833}/45\right) \approx 0.02429$ and $\mu_3 = \frac{1}{2}\left(1 - \sqrt{213}/15\right) \approx 0.01352$ (Exercise 4.40); therefore, these values of the μ parameter should be excluded.

The Hamiltonian $H_0 = H_0(I_1, I_2)$ is completely integrable and has a chance of satisfying the non-degeneracy conditions (4.35) and (4.36). It turns out that only condition (4.36) is significant. French astronomers A. Deprit and A. Deprit-Bartholomé proved that condition (4.36) is violated for one specific value of the parameter, $\mu_c = \frac{1}{2} - \left(2\sqrt{3}/9\right)\sqrt{3265/2576 + \sqrt{199945}/1288} \approx 0.0109$. Let us assume then that μ satisfies all the conditions listed above, which implies the thesis of the KAM theorem.

How does the stability result from the KAM theorem? Well, we are in a four-dimensional space near an equilibrium point. Because the system is Hamiltonian with the Hamiltonian independent of time, the movement takes place on the hypersurfaces $H = $ const. They are three-dimensional, like in Fig. 4.6. From the KAM theorem it follows that every such hypersurface is almost completely filled with the invariant tori, and the closer to the singularity $I_1 = I_2 = 0$ we are the more such invariant tori exist. Each invariant torus divides the hypersurface $H = \text{const}$ into two parts, its interior and exterior. No interior point comes out of it during the evolution. Because in the space of P, Q variables the tori can be arbitrarily small, this results in the stability in Lyapunov sense.

We will complete the above example. Let us assume that we have a Hamiltonian in the form

$$H = \sum \omega_j \cdot \frac{1}{2}(p_j^2 + q_j^2) + \cdots.$$

Definition 4.13. We say that the "frequencies" ω_j satisfy a **resonant relations of order** d, if there are integers $\alpha_1, \ldots, \alpha_n \in \mathbb{Z}$ with $\sum |\alpha_j| = d$ such that

$$(\alpha, \omega) = \sum \alpha_j \omega_j = 0.$$

Theorem 4.14 (Birkhoff). *If the frequencies ω_j do not satisfy any resonant relation of order $\leq 2m$, then there exists a canonical change of variables $(p, q) \longmapsto (P, Q) = (p + \cdots, q + \cdots)$ leading to the Hamiltonian*

$$H = \sum_{|\beta| \leq m} a_\beta I^\beta + O(|(p, q)|^{2m+1}),$$

where $I_j = \frac{1}{2}(P_j^2 + Q_j^2)$, the summation runs over multi-indices $(\beta_1, \ldots, \beta_n) \in (\mathbb{Z}_+)^m$ with $|\beta| = \beta_1 + \cdots + \beta_n$ and $I^\beta = I_1^{\beta_1}, \ldots, I_n^{\beta_n}$.

Proof. The change $(p, q) \longmapsto (P, Q)$ from the thesis of the Birkhoff theorem is **symplectic** (or *canonical*), i.e., it satisfies

$$\sum dp_j \wedge dq_j = \sum dP_j \wedge dQ_j.$$

It turns out that, after the canonical change of variables, a Hamiltonian system changes into a Hamiltonian system (see [11] and Definition 6.26 and Theorem 6.27 from Appendix).

The above change is symplectic when the 1-form $PdQ + qdp = \sum P_j dQ_j + \sum q_j dp_j$ is closed, and hence exact; it equals a differential of a function. So, there exists a so-called *generating function* $S(p, Q)$ such that $P = \partial S/\partial Q$ and $q = \partial S/\partial p$, where we assume that (p, Q) form a local coordinate system.

In our case, the function S expands into a Taylor series which begins with $(p, Q) = \sum p_j Q_j$. We have $S = (p, Q) + S_1$ and $P = p + \partial S_1/\partial Q$ (i.e., $p = P - \partial S_1/\partial Q$), $q = Q + \partial S_1/\partial p$.

We use the complex coordinates $z = p + iq$, $Z = P + iQ$, $i = \sqrt{-1}$. We get modulo higher order terms

$$z = Z + 2i\partial S_1/\partial \overline{Z}, \quad \bar{z} = \overline{Z} - 2i\partial S_1/\partial Z,$$

and $H = \sum \frac{1}{2}\omega_j z_j \bar{z}_j + H_1(z, \bar{z})$. Let us look how our change acts on H_1 when S_1 is a monomial,

$$S_1 = az^\alpha \bar{z}^\beta.$$

(Like in the proof of the Poincaré–Dulac theorem, our change is a composition of changes with homogeneous nonlinear parts, i.e., with homogeneous S_1's.) Then H_1 becomes increased by $i(\alpha - \beta, \omega) \cdot S_1 +$ h.o.t. We see that we can cancel all terms in H_1 but $|z_1|^{2k_1} \cdots |z_n|^{2k_n}$, $k_1 + \cdots + k_n < m$, provided there are no resonances of orders $\leq 2m$. □

Example 4.15 (The Lagrange top). Recall that in Example 4.5 we have presented two approaches to the dynamics of the rigid body in the Lagrange case.

One approach leads to a rather standard Hamiltonian system on the phase space $T^*SO(3)$, where $SO(3)$ (the group of three-dimensional rotations) is the configuration space of the system. Thus, it has three degrees of freedom with the Euler angles φ, ψ, θ as generalized coordinates. In the Lagrange case, the functions $H = T + U$ (the energy), $p_\varphi = \frac{\partial L}{\partial \varphi} = m_3$ (the vertical component of the angular momentum) and $p_\psi = \frac{\partial L}{\partial \psi} = M_3$ (the component of the angular momentum along the symmetry axis) are first integrals. The Liouville–Arnold tori $\mathbb{T}_{E,m_3,M_3}$ are three-dimensional and the energy levels $\{H = E\}$ are five-dimensional. Assume also that the non-degeneracy conditions (4.35)–(4.36) are satisfied.

Let us perturb slightly the parameters $I_1, I_2, I_3, K_1, K_2, K_3$ of the body (the principal moments of inertia and components of the center of mass position). Recall that before perturbation we had $I_1 = I_2$ and

$K_1 = K_2 = 0$; after perturbation these equations become slightly violated. By the KAM theorem most of the invariant tori survives. But this does not mean the stability of the motion, because these tori do not divide the hypersurfaces $\{H = E\}$ into invariant components; the differences $\{H = E\} \setminus torus$ are connected sets. One can expect that the action variables of a phase point can evolve far away from its initial values; this phenomenon is known as the *Arnold's diffusion*.

But we have another approach to the problem. It uses the Euler–Poisson equations (4.8) and the Poisson structure (4.9). Recall that this structure is degenerate with the symplectic leaves $\mathcal{Y} = \mathcal{Y}_{m_3} = \{\Gamma^2 = 1, (M, \Gamma) = -m_3\}$ and the Hamiltonian vector field $X_H = \{\cdot, H\}$ restricted to such leaf is a genuine Hamiltonian vector field with two degrees of freedom.

In the Lagrange case, the Liouville–Arnold tori $\mathbb{T}_{E,M_3} \subset \mathcal{Y}$ are two-dimensional and divide the energy levels into invariant components. By the KAM theorem, under non-degeneracy condition (4.36) (which we do not check) after violating slightly the Lagrange conditions, most of these tori survives and the movement of a phase point evolves either on such a torus or lies between invariant tori.

Therefore, the qualitative character of the motion is preserved. The top undergoes a precession and a nutation, although slightly perturbed.

We finish this section by some general remarks.

Important and still unsolved is the problem of stability of our Solar system. If the masses of the planets were negligible, then each planet would move along its elliptical orbit, independently on the other planets. So, the whole system would move on a torus. When we assume small masses, then this movement becomes perturbed.

Before the KAM theory only Fourier type expansions were used. But those series are generally divergent, because of terms of the form

$$\frac{1}{(k, \omega)} \exp(\mathrm{i}(k, \omega)t),$$

where $k \in \mathbb{Z}^n$, ω is the frequency vector and (k, ω) is the scalar product. One can assume that $(k, \omega) \neq 0$, but these quantities can be arbitrarily small. In the proof of the KAM theorem, one controls these series in the cases when (k, ω) do not approach zero too quickly, one assumes $|(k, \omega)| > C |k|^{-\nu}$ for some constants C and ν.

However, even the existence of the invariant tori does not guarantee the stability, due to the so-called Arnold's diffusion. If the number degrees of freedom exceeds 2, then the action variables can evolve very far from their initial values. That evolution is very small, but inevitable.

Another question concerns the dynamics between the invariant tori. It can be irregular.

For example, in the case of two degrees of freedom the resonant tori degenerate after perturbation. There appear periodic solutions (which can be detected using suitable Melnikov functions like in the case of subharmonic solutions from Section 4.1.4). Half of these solutions is of the saddle type and the other half is of the elliptic type.

To explain this, consider corresponding Poincaré return map f on a fixed energy level. It is a two-dimensional map, which preserves some measure absolutely continuous with respect to the Lebesque measure; note that any Hamiltonian vector field preserves the Liouville measure (see the next chapter). Therefore, if z_0 is a periodic point of f of period p (corresponding to a closed phase curve), then

$$\det Df^p(z_0) = 1.$$

So, the eigenvalues satisfy $\lambda_1 \lambda_2 = 1$. If they are real and $\neq 1$, then we have a hyperbolic periodic point of the saddle type. Otherwise, generically, the eigenvalues are conjugate complex numbers of module 1; here we have an *elliptic periodic point*, which can be Lyapunov stable or unstable.

Next, the saddle type periodic points have invariant manifolds which generally intersect transversally other invariant manifolds. This leads to a chaotic behavior of orbits, as described in the next chapter.

Moreover, near an elliptic periodic orbit of a period T and the multipliers $\lambda_{1,2} = e^{\pm 2\pi i \alpha}$ with irrational α there exist invariant thin tori around it (again due to the KAM theorem), but between such tori there usually exist periodic orbits of very long period NT; again half of them of the saddle type and the other half of the elliptic type. Also here a chaotic dynamics is predicted, but is hard to investigate.

4.3. Partially integrable systems

By a **partially integrable system** we mean a differential autonomous system with an invariant submanifold which is filled with periodic or quasi-periodic solutions. Moreover, such invariant submanifold is isolated (in contrast to the completely integrable Hamiltonian systems).

The simplest such situation is a three-dimensional system with a plane supporting a Hamiltonian system. The corresponding equations are as follows:

$$\dot{x} = H'_y + zR, \quad \dot{y} = -H'_x + zS, \quad \dot{z} = az, \qquad (4.39)$$

where $H(x, y)$ (the Hamiltonian), $R(x, y, z)$, $S(x, y, z)$ and $a(x, y, z)$ are some functions. Here the invariant manifold is $S = \{z = 0\}$. We assume that some some level curves $\{H(x, y) = h\}$, $h_1 < h < h_2$, contain compact ovals γ_h.

We want to study what happens with the invariant plane $\{z = 0\}$ and with the periodic orbits γ_h when we add a perturbation to the right and sided , i.e., for the system

$$\dot{x} = H'_y + zR + \varepsilon P, \quad \dot{y} = -H'_x + zS + \varepsilon Q, \quad \dot{z} = a + \varepsilon b, \qquad (4.40)$$

where $P(x, y, z)$, $Q(x, y, z)$ and $b(x, y, z)$ are some functions.

We have the following notion from the Dynamical Systems Theory.

Definition 4.16. Let $f : M \longmapsto M$ be a diffeomorphism of a differentiable manifold and $L \subset M$ be an invariant submanifold, $f(L) \subset L$. We say that f is **normally hyperbolic** on L if there is a splitting of the normal to L bundle $NL = TM|_L/TL$, into a Whitney sum

$$NL = E^u \oplus E^s,$$

which is preserved by Df and:

- the bundle E^u is expanded more strongly than the tangent bundle TL, i.e.,

$$\inf_L \| (Df|_{E^u})^{-1} \| > \sup_L \|Df|_{TL}\|,$$

- the bundle E^s is contracted more strongly that the tangent bundle, i.e.,

$$\sup_L \|Df|_{E^s}\| < \inf_L \| (Df|_L)^{-1} \|.$$

If v is a vector field on M tangent to a submanifold S, then we say that it is **normally hyperbolic** on S if the flow map g_v^T (after some time $T > 0$) is normally hyperbolic to S.

We have the following result which we present without proof.

Theorem 4.17 (Hirsch, Pugh, Shub). *Assume a situation like in the above definition. Then there exists a neighborhood of f in a functional space of C^1-diffeomorphisms of M such that any diffeomorphism f_ε from this neighborhood has an invariant submanifold L_ε, close to L, and f_ε is normally hyperbolic on L_ε.*

In the context of system (4.39), we take a two-dimensional section $M \subset \mathbb{R}^3$ transversal to the closed orbits $\gamma_h \subset \{z = 0\}$ and the Poincaré return map $f : M \longmapsto M$ with the invariant line $L = M \cap \{z = 0\}$.

Let $(x, y) = \phi_h(t)$ be the periodic solution (of period $T(h)$) of the Hamiltonian system on $\{z = 0\}$ corresponding to the closed phase curve γ_h and with the initial condition on M.

We introduce local coordinates (h, z) on M such that $h|_L = H|_L$. In these coordinates, the return map takes the form

$$f(h, z) = \left(h + O(z), \lambda z + O(z^2) \right),$$

where $\lambda = e^\mu > 0$ with the **Lyapunov exponent**

$$\mu = \mu(h) = \int_0^{T(h)} a(\phi_h(t)) \, \mathrm{d}t. \tag{4.41}$$

Since the restriction of the map f to L is the identity, then this map is normally hyperbolic to L when the Lyapunov exponent $\mu(h) \neq 0$. If $\mu < 0$ then the plane $S_0 = \{z = 0\}$ is attracting, if $\mu > 0$, then it is repelling.

Let us perturb the above situation, assuming $\mu(h) \neq 0$ for h's from some interval $[h_1, h_2]$. Thus, we have system (4.40). We expect an invariant surface S_ε of the form

$$z = \varepsilon g(x, y) + O(\varepsilon^2). \tag{4.42}$$

Denote $\tilde{a}(t) = \tilde{a}_h(t) = a(\phi_h(t), 0)$, $\tilde{b}(t) = \tilde{b}_h(t) = b(\phi_h(t), 0)$, $\tilde{g}(t) = \tilde{g}_h(t) = g(\phi_h(t))$.

Lemma 4.18. *The linear approximation g is such that $\tilde{g}_h(t)$ is the unique periodic solution to the following **normal variations equation**:*

$$\frac{d}{dt}\tilde{g} = \tilde{a}(t)\tilde{g} + \tilde{b}(t). \tag{4.43}$$

Proof. The invariance of a surface $z = G(x, y) = \varepsilon g + \cdots$ means $aG + \varepsilon b \equiv G'_x \left(H'_y + \cdots \right) + G'_y \left(-H'_x + \cdots \right)$ on this surface. This leads to the PDE $H'_y g'_x - H'_x g'_y = (ag + b)|_{z=0}$, which is equivalent to Eq. (4.43).

Let

$$A(t) = A_h(t) = \int_0^t \tilde{a}_h(s)\, ds$$

and $T = T(h)$ be the period of the solution $\phi_h(t)$. Then the unique periodic solution to Eq. (4.43) is

$$\tilde{g}(t) = \tilde{g}_h(t) = \frac{1}{1 - e^{\mu(h)}} \int_{t-T}^t e^{A(t)-A(s)} \tilde{b}(s)\, ds \tag{4.44}$$

(Exercise 4.42). $\qquad\qquad\qquad\qquad\qquad\qquad\qquad\qquad\qquad\square$

After restriction of system (4.40) to the invariant surface S_ε (parametrized by x, y) we get a perturbation of a Hamiltonian system

$$\dot{x} = H'_y + \varepsilon(P + gR) + \cdots, \quad \dot{y} = -H'_x + \varepsilon(Q + gS) + \cdots,$$

like Eq. (4.21) in Section 4.1.3, and the condition for limit cycles takes the form $I(h) + O(\varepsilon) = 0$, where the Melnikov function I equals

$$I(h) = \oint_{\gamma_h} (Q + gS)\, dx - (P + gR)\, dy,$$

where $P = P(x, y, 0)$, etc. Let us state the following analogue of Proposition 4.7.

Proposition 4.19. *If $I(h_0) = 0$ and $I'(h_0) \neq 0$, then for small $\varepsilon \neq 0$ there exists a limit cycle γ_ε which tends to the curve γ_{h_0} as $\varepsilon \to 0$.*

Above the part $\oint Q\mathrm{d}x - P\mathrm{d}y$ is the standard Abelian integral (4.22). But the other part $J(h) = \oint g\,(S\mathrm{d}x - R\mathrm{d}y)$ is of new type. Using the above integral formula for g, we can express it in the form of a double integral

$$J(h) = \frac{1}{1 - e^{\mu(h)}} \int_0^T \int_{t-T}^t e^{A(t)-A(s)} \tilde{b}(s)\tilde{c}(t)\mathrm{d}s\mathrm{d}t, \qquad (4.45)$$

where

$$\tilde{c}(t) = \left[SH_y' + RH_x'\right](\phi_h(t), 0).$$

Remark 4.20. One can generalize the above to the case when the variable z is of higher dimension. Then $a = a(x, y, z)$ is a matrix-valued function, b, R, S are vector-valued functions and $zR = (z, R)$ and $zS = (z, S)$ become scalar products.

Instead of the normal multiplier $\lambda = e^{\mu(h)}$ we have the *monodromy matrix*

$$\mathcal{M}_h = \mathcal{F}(T(h), 0), \qquad (4.46)$$

where $\mathcal{F}(t, s)$ is the *fundamental matrix* of the linear homogeneous system

$$\dot{z} = \tilde{a}(t)\, z$$

such that $\mathcal{F}(s, s) = 0$; it is the 2-parameter family defining the evolution of the latter linear system.

The normal hyperbolicity condition is satisfied when the matrix $\mathcal{M}_h$ is hyperbolic, i.e., its eigenvalues satisfy $|\lambda_j| \neq 1$.

The generalization of the formula (4.44) for the linear term in the invariant surface equation is as follows:

$$\tilde{g}(t) = (I - \mathcal{M}_h)^{-1} \int_{t-T}^T \mathcal{F}(t, s)\, \tilde{b}(s)\mathrm{d}s.$$

Finally, the double integral (4.45) becomes

$$J(h) = \int_0^T \int_{t-T}^t ((I - \mathcal{M})^{-1} \mathcal{F}(t, s)\, \tilde{b}(s), \tilde{c}(t))\mathrm{d}s\mathrm{d}t.$$

Generalized Melnikov Functions as above were investigated by M. Bobieński, P. Leszczyński and the first author in particular cases. Some estimates for the number of limit cycles which arise in the perturbed invariant surface were obtained.

But one cannot expect that always the number of such limit cycles is finite. A four-dimensional example with infinite number of limit cycles was found.[c]

In the following, we present some systems from applied mathematics with the partial integrability property.

Example 4.21 (Three-dimensional generalized Lotka–Volterra systems). These are the following systems:

$$\dot{x} = x\,(ay + bz + \lambda u), \quad \dot{y} = y\,(cz + dx + \mu u), \quad \dot{z} = z\,(ex + fy + \nu u),$$
(4.47)

where

$$u = z + y + z - 1.$$

We denote this vector field by X.

Four cases of partial integrability were found. We present them here:

(1) This case is characterized by the conditions

$$a + d = b + e = c + f = 0$$

onto the coefficients in Eq. (4.47). Under them, the plane

$$S = \{u = 0\}$$

becomes invariant; we get $\dot{u} = (\lambda x + \mu y + \nu z)\,u$. This plane supports a two-dimensional Lotka–Volterra system $X|_S$, like in Example 2.38 from Chapter 2, with a first integral of the Darboux form

$$H = |x|^\alpha\,|y|^\beta\,|z|^\gamma,$$

restricted to S.

---

[c]See, for example, M. Bobieński and H. Żołądek, Limit cycles for three-dimensional vector fields, *Nonlinearity* 18 (2005), 175–209 and M. Bobieński and H. Żołądek, A counterexample to a multidimensional version of the weakened Hilbert's 16th problem, *Moscow Math. J.* 7 (2007), 1–20.

(2) Under the conditions

$$e + \lambda = f + \nu = b\mu(\lambda - \mu - d) + c\lambda(\mu - \lambda - a) = 0,$$

the plane

$$S = \{z = 1\}$$

becomes invariant, $\dot{z} = \nu z \, (z - 1)$. The restricted vector field X_S is again a two-dimensional Lotka–Volterra system with a first integral of the Darboux form (product of powers of linear functions).

(3) Under the conditions

$$a + \lambda = c + \mu = a + d = 0,$$

the vector field X has the Darboux function

$$F = |x|^\alpha \, |y|^\beta \, |1 - x - y|^\gamma$$

as a first integral (thus the plane $x + y = 1$ is invariant, but this is not the invariant surface we are interested in). Note that the function F depends on two variables and has a line $\ell = \{x = x_*, y = y_*\}$ of critical points; under some additional conditions F has a maximum along ℓ. This line is invariant for X and contains a singular point p_*, for which ℓ is either a stable manifold or an unstable manifold. But it has also its two-dimensional center manifold, which is our invariant surface,

$$S = W^c \, (p_*).$$

The closed phase curves on S are $\{F = \text{const}\} \cap S$.

(4) The last case is defined by the following conditions:

$$\begin{vmatrix} 0 & a & b \\ d & 0 & c \\ e & f & 0 \end{vmatrix} = df\lambda + ae\mu + ad\nu = 0.$$

They imply, on the one hand, that there exists a line ℓ of non-isolated equilibrium points for X, and, on the other hand, that the function

$$E = |x|^\alpha \, |y|^\beta \, |z|^\gamma,$$

$\alpha = df$, $\beta = ae$, γad, is a global first integral for X. So, we have a 1-parameter family $v_\kappa = X|_{\{E = \kappa\}}$ of two-dimensional vector fields with singular points $p_\kappa = \ell \cap \{E = \kappa\}$. Under some additional conditions (inequalities), an Andronov–Hopf bifurcation takes place for a bifurcational value κ_* of the parameter. The family of limit cycles created in this bifurcation cover a surface S on which X is partially integrable.

Example 4.22 (Hess–Appelrot case in the rigid body dynamics). The **Hess–Appelrot case** is the Euler–Poisson system (4.8), i.e.,

$$\dot{M} = M \times \Omega - \Gamma \times K, \quad \dot{\Gamma} = \Gamma \times \Omega$$

with the following restrictions:

$$K_2 = 0, \quad K_1 \sqrt{I_1 (I_2 - I_3)} = K_3 \sqrt{I_3 (I_1 - I_2)}, \tag{4.48}$$

where $0 < I_1 < I_2 < I_3$ and $K_1 K_3 \neq 0$. The second of condition (4.48) can be rewritten in the form

$$\frac{I_2 - I_3}{I_3 K_3^2} = \frac{I_1 - I_2}{I_1 K_1^2}.$$

Introduce the following variables:

$$X = (M, M), \quad Y = (M, K \times \Gamma), \quad Z = (M, K). \tag{4.49}$$

Recall that $M = I\Omega$, where Ω is the angular velocity, M is the angular momentum, K is a constant vector parallel to the center of mass position and Γ is the unit vector in the direction of the gravity force. All vectors are in the coordinate system associated with the body.

It follows directly from the Euler–Poisson equations (4.8) that

$$\dot{Z} = 2Y.$$

Next, under the Hess–Appelrot conditions, we have

$$\dot{Z} = \eta \Omega_2 \cdot Z,$$

$\eta = K_1 (I_2 - I_3) / K_3 I_1$. Therefore, the so-called **Hess surface**

$$S = \{Z = 0\} \tag{4.50}$$

is invariant with respect to the vector field. Recall that we have a Hamiltonian vector field X_H, $H = \frac{1}{2} (M, \Omega) - (K, \Gamma)$, restricted to the four-dimensional symplectic leaf $\mathcal{Y} = \mathcal{Y}_{m_3} = \{\Gamma^2 = 1, \ (M,\Gamma) = -m_3\}$. The condition $z = 0$, when restricted to the three-dimensional energy level $\{H = E\} \subset Y$ defines a two-dimensional manifold.

Introduce the following function on the manifold $\mathcal{Y}$:

$$F := (M,\Omega)\,/M^2 = (I\Omega,\Omega)\,/X,$$

where (M,Ω) is the double kinetic energy. It turns out that this function restricted to the Hess surface S is constant,

$$F|_S \equiv 1/I_2.$$

To prove this, we note the equations

$$(I\Omega,\Omega) = Fx, \quad (I\Omega, I\Omega) = X, \quad I_1^2 K_1^2 \Omega_1^2 - I_3^2 K_3^2 \Omega_3^2 = 0,$$

which lead to the linear system

$$\begin{pmatrix} I_1 & I_2 & I_3 \\ I_1^2 & I_2^2 & I_3^2 \\ I_1^2 K_1^2 & 0 & -I_3^2 K_3^2 \end{pmatrix} \begin{pmatrix} \Omega_1^2 \\ \Omega_2^2 \\ \Omega_3^2 \end{pmatrix} = \begin{pmatrix} FX \\ X \\ 0 \end{pmatrix},$$

where the determinant of the above matrix is zero due to the second of the Hess–Appelrot conditions. It is easy to check that the solvability conditions for the above system are: $X = 0$ (but then we arrive to the non-interesting case $\Omega = 0$) and $F = 1/I_2$.

We represent the vector Γ in the basis $(M, K, M \times K)$,

$$\Gamma = \alpha \cdot M + \beta \cdot K + \gamma \cdot M \times K.$$

Since $\alpha x = (M,\Gamma) = -m_3$, $\beta K^2 = (K,\Gamma) = X/(2I_2) - E$, $\gamma K^2 X = (I\Omega \times K, \Gamma) = Y$, we find

$$\alpha = -m_3/X, \quad \beta = (X/2I_2 - E)\,/K^2, \quad \gamma = Y/K^2 X.$$

Then the geometrical condition $\Gamma^2 = 1$ leads to the equation

$$Y^2 = R(X), \tag{4.51}$$

where $R = -X\,(X/2I_2 - E)^2 + K^2 X - (m_3 K)^2$. Equation (4.51) is an equation for an elliptic curve in the (X, Y)-plane. Usually such curve has one or two components, from which one is unbounded; the bounded component (oval) γ is diffeomorphic to a circle (or it is a point) and lies in the domain $\{Y > 0\}$.

When expressing $\Omega_3|_S$ via Ω_1 we arrive to the following formula:

$$X = (I_1 |K| / K_3)^2 \cdot \Omega_1^2 + I_2^2 \cdot \Omega_2^2 = J_1^2 \Omega_1^2 + J_2^2 \Omega_2^2. \qquad (4.52)$$

From the above, it follows that the Hess surface (4.50) is a torus. Indeed, we have the map

$$(\Omega, \Gamma) \longmapsto (\Omega_1, \Omega_2, Y),$$

which is a diffeomorphism from S to its image (define by Eqs. (4.51)–(4.52)) which is a torus (like $\gamma \times \mathbb{S}^1$) in $\mathbb{R}^3$.

Using the representation of Γ in the our orthogonal frame, the first triple of the Euler–Poisson equations reads as $I\dot{\Omega} = I\Omega \times \Omega + \frac{1}{X}\{m_3 \cdot I\Omega \times K + Y \cdot I\Omega\}$. It gives the following system for the first two components of Ω:

$$\dot{\Omega}_1 = -J_2^2(\tilde{a}\Omega_1 + \tilde{b}/X)\Omega_2 + (Y/X)\Omega_1,$$
$$\dot{\Omega}_2 = J_1^2(\tilde{a}\Omega_1 + \tilde{b}/X)\Omega_1 + (Y/X)\Omega_2,$$

where $\tilde{a} = (I_2 - I_3)K_1/I_2^2 I_3 K_3 > 0$ and $\tilde{b} = -m_3 K_3 / I_1 I_2$.

Before passing to more detailed analysis of the above equations we simplify them by making some normalizations. We put

$$X = 2I_2 |K| \cdot x, \quad Y = \sqrt{2I_2 |K|^3} \cdot y,$$

$$\Omega_{1,2} = (\sqrt{2I_2 |K|}/J_{1,2}) \cdot \omega_{1,2}, \quad Z = z$$

(where $J_{1,2}$ are defined above). We get the following equations:

$$\begin{aligned}
\dot{x} &= 2gy + O(z), \\
y^2 &= R(x) = -x(x - c)^2 + x - H + O(z), \\
\dot{\omega}_1 &= -2(a\omega_1 + b/x)\omega_2 + g(y/x)\omega_1 + O(z), \qquad (4.53) \\
\dot{\omega}_2 &= 2(a\omega_1 + b/x)\omega_1 + g(y/x)\omega_2 + O(z), \\
\dot{z} &= f\omega_2 z,
\end{aligned}$$

where

$$\omega_1^2 + \omega_2^2 = x,$$

and $a > 0$, b, c, $\kappa > 0$, $g > 0$ and $H \geq 0$ are constants. The system (4.53) is called the *Hess–Appelrot system*.

We can simplify the equations for ω_j in the system (4.53). Let

$$\varphi = \arg(\omega_1 + i\omega_2), \quad u = \tan(\varphi/2),$$

$i = \sqrt{-1}$; thus $\omega_1 = \sqrt{x}\cos\varphi$ and $\omega_2 = \sqrt{x}\sin\varphi$. Then from the above equations we get

$$\dot{u} = \frac{\dot{\varphi}}{2\cos^2(\varphi/2)} = \frac{1}{2\cos^2(\varphi/2)} \cdot \frac{\dot{\omega}_2\omega_1 - \dot{\omega}_1\omega_2}{x}$$

$$= \frac{1}{\cos^2(\varphi/2)} \cdot \left\{ a\sqrt{x}\cos\varphi + (b/x)\cdot 1 \right\}$$

$$= \left\{ b/x + a\sqrt{x} \right\} + \left\{ b/x - a\sqrt{x} \right\} u^2.$$

We arrive at the following differential–algebraic system:

$$\dot{u} = A(x) + B(x)u^2,$$

$$\dot{x} = \pm 2g\sqrt{R(x)}, \qquad\qquad (4.54)$$

$$A = b/x + a\sqrt{x}, \quad B = b/x - a\sqrt{x}.$$

The second of the above equations is solved by means of one elliptic function,

$$x = \mathcal{P}(t),$$

where $\mathcal{P}(t)$ is the Weierstrass $\mathcal{P}$-function (compare analogous situation in the Lagrange case in Example 4.5). The function $\mathcal{P}(t)$ is periodic with the period

$$T_1 = \frac{1}{2g} \oint_\gamma \frac{dx}{y},$$

where the integral runs along the oval $\gamma = \left\{ y^2 = R(x),\ x > 0 \right\} \subset \mathbb{R}^2$ (see Fig. 4.7).

Finally, we arrive at the following *Riccati equation* with periodic coefficients:

$$\dot{u} = \Theta(t) + \Xi(t)u^2, \qquad\qquad (4.55)$$

$\Theta = A \circ \mathcal{P}(t), \quad \Xi = B \circ \mathcal{P}(t).$

The Riccati form of Eq. (4.55) for $u = \tan\frac{1}{2}\arg(\omega_1 + i\omega_2)$ is very special. Namely, we can assume that $u \in \mathbb{RP}^1 = \mathbb{R} \cup \infty$ and

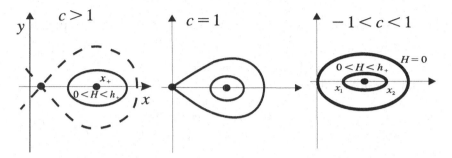

Fig. 4.7. The elliptic ovals.

Eq. (4.55) defines a regular vector field at $\mathbb{RP}^1$. Indeed, in the chart $v = 1/u$ near ∞ we have $\dot{v} = -\Theta(t)v^2 - \Xi(t)$. Therefore, the evolution maps $u_0 \longmapsto g_s^t(u_0)$, i.e. solution $u(t)$ at the time t with the initial condition $u(s) = u_0$ at time s, are automorphisms of $\mathbb{RP}^1$ (they are also automorphisms of $\mathbb{CP}^1$). Hence, they are the Möbius maps,

$$g_s^t(u) = \frac{\alpha(s,t)u + \beta(s,t)}{\gamma(s,t)u + \delta(s,t)}.$$

Important is the *monodromy map*

$$\mathcal{M} = g_0^T = \frac{\alpha u + \beta}{\gamma u + \delta} : \mathbb{RP}^1 \longmapsto \mathbb{RP}^1. \tag{4.56}$$

It is defined by means of the matrix

$$\mathcal{A} = \begin{pmatrix} \alpha & \beta \\ \gamma & \delta \end{pmatrix} \in \mathrm{SL}(2,\mathbb{R}),$$

where we can assume that $\det \mathcal{A} = 1$, because g_0^T is connected with $g_0^0 = id$ and the matrices $\mathcal{A}$ and $\lambda\mathcal{A}$ give the same monodromy maps.

All the dynamics and geometry of the Hess–Appelrot system on the invariant surface $\mathcal{T}$ is completely determined by means of the map (4.56).

Let us recall the standard classification of the Möbius maps defined by means of $\mathcal{A} \in \mathrm{SL}(2,\mathbb{R})$:

(a) $|\mathrm{tr}\,\mathcal{A}| > 2$, i.e., the matrix $\mathcal{A}$ is equivalent to $\begin{pmatrix} \lambda & 0 \\ 0 & \lambda^{-1} \end{pmatrix}$, $\lambda > 1$;

here the map $\mathcal{M}$ is *hyperbolic* with two fixed points u_1 and u_2 (one attracting and one repelling) in $\mathbb{RP}^1$.

(b) $\mathcal{A}$ is equivalent to $\begin{pmatrix} 1 & 1 \\ 0 & 1 \end{pmatrix}$, thus $|\operatorname{tr}\mathcal{A}| = 2$ and $\mathcal{M}$ is conjugated with translation $u \longmapsto u + 1$. Here $\mathcal{M}$ is *parabolic* with only one (non-hyperbolic) fixed point $u_0 \in \mathbb{RP}^1$.

(c) $\mathcal{A}$ is equivalent to

$$\begin{pmatrix} \cos(2\pi\vartheta) & \sin(2\pi\vartheta) \\ -\sin(2\pi\vartheta) & \cos(2\pi\vartheta) \end{pmatrix}, \tag{4.57}$$

thus $|\operatorname{tr}\mathcal{A}| \leq 2$ and $|\operatorname{tr}\mathcal{A}| = 2$ only when $\mathcal{M} = id$. Here $\mathcal{M}$ is *elliptic* and, in the chart $\phi = \tan^{-1} u$ for $\mathcal{A}$ of the form (4.57), it takes the form

$$\phi \longmapsto \phi + 2\pi\vartheta \quad \mathrm{mod}\ 2\pi.$$

For ϑ irrational we call the map $\mathcal{M}$ *elliptic irrational* and for

$$\vartheta = p/q \in \mathbb{Q}$$

(reduced ratio with $p, q \in \mathbb{Z}$, $q > 0$) we say that $\mathcal{M}$ is *elliptic rational*.

If the monodromy map is elliptic irrational, then all its orbits are dense in $\mathbb{RP}^1$. But for an elliptic rational map all orbits of $\mathcal{M}$ are periodic with period q. (An elliptic monodromy map acting on $\mathbb{CP}^1$ has also two fixed points, but they are not real.)

It is the elliptic case of dynamics on the invariant Hess surface that we have a partially integrable system.

Next step in the analysis of the above situation is to consider perturbations of the Hess–Appelrot case with partial integrability. Such perturbations and arising limit cycles were studied by R. Kurek, P. Lubowiecki and the first author in two limit situations[d]:

(i) when the oval γ is approximately a small circle around a critical point, and

(ii) when the oval γ is closet to a loop with vertex at $x = y = 0$ (here a chaotic behavior is observed).

---

[d]See P. Lubowiecki and H. Żołądek, The Hess-Appelrot system. II. Petrurbation and limit cycles, *J. Differential Equations* 252 (2012), 1701–1722.

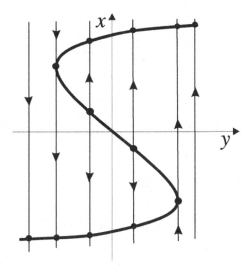

Fig. 4.8. van der Pol slow–fast system.

4.4. Slow–fast systems

Let us start with a known example.

Example 4.23 (van der Pol system). Consider the system

$$\dot{x} = y - x^3 + x, \quad \dot{y} = -\varepsilon x.$$

(When $\varepsilon = 1$ then, putting $y_1 = y - x^3 + x$, one gets $\dot{x} = y_1$, $\dot{y}_1 = -x - (3x^2 - 1)y_1$ which, up to a scaling, is the system from Example 2.39 in Chapter 2).

You can see that x varies quickly compared to y; we say that x is a **fast variable** and y is a **slow variable**. For $\varepsilon = 0$ we have $y(t) \equiv$ const and, in fact, we have an equation for x depending on the parameter y (Bifurcation Theory bows, see Fig. 4.8). When $\varepsilon \neq 0$ (but small) physicists would say that the parameter y "flows". It is expected that there will be a limit cycle γ_ε (in fact γ_ε is stable) tending to a piecewise smooth curve γ_0 shown in Fig. 4.9. The γ_0 cycle consists of:

- pieces of slow motion along the curve $y = x^3 - x$ (where $\dot{x} = 0$);
- segments of jump along the straight lines $y =$ const.

Such movement is an example of **relaxation oscillations** (like a heartbeat).

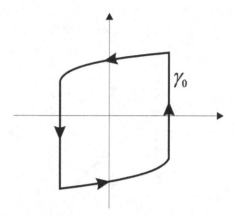

Fig. 4.9.   Relaxational oscillations.

Consider now the general situation. We have an **unperturbed system**

$$\dot{x} = f(x, y), \quad \dot{y} = 0$$

($x \in \mathbb{R}^k$, $y \in \mathbb{R}^l$); here x are the **fast coordinates** and y are the **slow coordinates**. We also have a **perturbed system**

$$\dot{x} = F(x, y; \varepsilon), \quad \dot{y} = \varepsilon G(x, y; \varepsilon), \quad F(x, y; 0) = f(x, y).$$

Definition 4.24. The surface $S = \{f(x, y) = 0\}$ is called the **slow surface**.

The slow surface is divided into **regions of stability** and **instability** of the unperturbed system; they correspond to situations when $\operatorname{Re} \lambda_j(A) < 0$, $j = 1, \ldots, k$, $A = \frac{\partial f}{\partial x}$, and when there exists $\operatorname{Re} \lambda_j(A) > 0$, respectively.

On a slow surface we have a vector field defined as follows. We take the vector

$$\frac{\partial}{\partial \varepsilon} \left(F \frac{\partial}{\partial x} + \varepsilon G \frac{\partial}{\partial y} \right) \Big|_{\varepsilon = 0} = f_1(x, y) \frac{\partial}{\partial x} + g(x, y) \frac{\partial}{\partial y}$$

at a point $(x, y) \in S$ and project it to $T_{(x,y)}S$ along the variable y. This is the **slow motion vector field**.

Let us remind that, at the beginning of this chapter, we were saying that relaxation oscillations are characterized by the fact that the small parameter appears on the left side. To see this, we introduce the **slow time** $\tau = \varepsilon t$. Then we get the system

$$\varepsilon \frac{dx}{d\tau} = f(x, y) + O(\varepsilon), \quad \frac{dy}{d\tau} = g(x, y) + O(\varepsilon).$$

Now, the **slow motion equation** to S (locally parameterized by y) has the form

$$\frac{dy}{d\tau} = h(y) + O(\varepsilon)$$

(with a corresponding function h).

Let us investigate the movement of a typical point (x_0, y_0). It consists of pieces of three types: approaching to the slow surface, movement along the slow surface and movement in a transition region.

4.4.1. *Approaching the slow surface*

Let the point (x_0, y_0) from outside S be projected (along the y coordinates) to the point $(x_*, y_0) \in S$, $x_* = x_*(y_0)$, in the stability region (see Fig. 4.10). This means that the point x_0 lies in the basin of attraction of the point x_* for the equation $\dot{x} = f(x, y_0)$ (y_0 fixed). Consider the domain $U = \{|x - x_*(y_0)| < \delta, \; y_0 \in V\}$, where V is a certain domain corresponding to a subset of the stability region in S.

It turns out that the slow time τ_1 needed to reach the domain U by the solution with the initial condition (x_0, y_0) is of order $\sim C_1 \varepsilon |\ln \varepsilon|$,

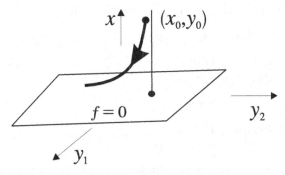

Fig. 4.10. Approaching slow surface.

which corresponds to the actual time

$$t_1 \sim C_1 \left| \ln \varepsilon \right|$$

(the constant C_1 depends on U and F, G).

4.4.2.   *Movement along the slow surface*

In the domain U, we have slow motion, described by the equation $dy/d\tau = h(y) + O(\varepsilon)$. It lasts until $\tau_2 = T = O(1)$, which corresponds to the long real time $t_2 = T/\varepsilon$.

4.4.3.   *Movement in the transition region*

The transition region lies close to the border between the regions of stability and instability in S. We have two typical options (like in the bifurcation theory):

A. $\lambda_1(A) = 0$ (where $A = \frac{\partial f}{\partial x} \big|_{f=0}$);
B. $\operatorname{Re} \lambda_{1,2}(A) = 0$.

A. Spurt.   This case, which corresponds to the saddle-node bifurcation, is analyzed for situations when $x \in \mathbb{R}$ and $y \in \mathbb{R}$ (you can reduce everything to this case). After appropriate rescaling, we have the following system:

$$\dot{x} = x^2 - y + \cdots, \quad \dot{y} = -\varepsilon + \cdots.$$

We perform the normalization

$$\varepsilon = \mu^3, \quad x = \mu X, \quad y = \mu^2 Y,$$

and it is easy to check that it leads to the field

$$\dot{X} = \mu \left\{ X^2 - Y + O(\mu) \right\}, \quad \dot{Y} = \mu \left\{ -1 + O(\mu) \right\}$$

orbitally equivalent to field $\left(X^2 - Y \right) \frac{\partial}{\partial X} - \frac{\partial}{\partial Y}$. Its phase portrait is given by the *Riccati equation*[e]

$$dX/dY = Y - X^2 \tag{4.58}$$

and is shown in Fig. 4.11.

---

[e]Equation (4.58) is probably the simplest example of a differential equation which cannot be solved in the so-called quadratures.

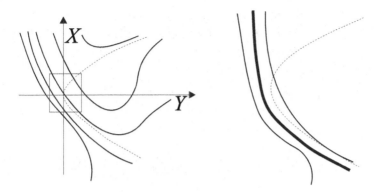

Fig. 4.11. Spurt.

The phenomenon which we observe here is called the **spurt**.

B. Delay of the loss of stability. In this case, which corresponds to the Andronov–Hopf bifurcation, the problem is reduced to the following model system:

$$\dot{z} = (y + i\omega)\,z + cz\,|z|^2, \quad \dot{y} = \varepsilon, \qquad (4.59)$$

$z = x_1 + ix_2 \in \mathbb{C} \simeq \mathbb{R}^2$, $y \in \mathbb{R}$. Of course, $y = \varepsilon t$ is a "flowing" parameter. Suppose further that

$$c = -1;$$

the case of $c > 0$ is less interesting. For the amplitude $r = |z|$, we get the Bernoulli equation

$$\dot{r} = r\left(\varepsilon t - r^2\right).$$

Let us put the initial condition

$$y(t_0) = -\mu, \quad r(t_0) = r_0, \quad t_0 = -\mu/\varepsilon,$$

where $\mu > 0$ is a fixed constant (not too big and not too small). This initial value problem has the following solution:

$$r(t) = r_0 \left\{ e^{\varepsilon(t_0^2 - t^2)} + 2r_0^2 \int_{t_0}^{t} e^{\varepsilon(s^2 - t^2)}\,\mathrm{d}s \right\}^{-1/2} \qquad (4.60)$$

(Exercise 4.43). We will examine the asymptotic behavior of this solution as $\varepsilon \to 0$ by dividing the range of time into four domains:

(1) $\{0 < t - t_0 < O(1)\}$, i.e., $0 < y + \mu < O(\varepsilon)$.
Let $u = t - t_0$. Then $\varepsilon\left(t_0^2 - t^2\right) = \varepsilon(t_0 + t)u \approx 2\mu u$ and $\varepsilon\left(s^2 - t^2\right) \approx 2\mu(u - v)$, where $v = s - t_0$. Thus,

$$\int_{t_0}^{t} e^{\varepsilon\left(s^2 - t^2\right)} ds \approx \int_0^u e^{2\mu(u-v)} dv = \frac{1}{2\mu}(e^{2\mu u} - 1)$$

and

$$r(t) \approx r_0 \left\{ e^{2\mu u} + r_0^2(e^{2\mu u} - 1)/\mu \right\}^{-1/2}$$

is a decreasing function of u.

(2) $y = \varepsilon t$ is fixed such that $-\mu < y < \mu$, i.e., $\{-\mu/\varepsilon < t < \mu/\varepsilon\}$.
Here $e^{\varepsilon\left(t_0^2 - t^2\right)} \approx e^{(\mu^2 - y^2)/\varepsilon} \to \infty$. Thus,

$$r(t) < C_1 e^{-C_2/\varepsilon} \to 0$$

which means very fast approaching zero.

(3) $\{0 < |t_0| - t < O(1)\}$, i.e., $0 < \mu - y < O(\varepsilon)$.
Let us introduce the variable $w = |t_0| - t$. As in case (1), we have $e^{\varepsilon\left(t_0^2 - t^2\right)} \approx e^{2\mu w}$.

We will divide the integration domain for the integral in Eq. (4.60) into three intervals: from $t_0 < 0$ to $t_0/2 < 0$, from $t_0/2 < 0$ to $|t_0|/2 > 0$ and from $|t_0|/2$ to t. Denote the corresponding integrals by I_1, I_2 and I_3, respectively. Similarly as in case (1), one shows that $I_1 = O(1)$ and $I_3 = O(1)$. The calculations in case (2) show that $I_2 \to 0$ very quickly. Thus,

$$r(t) = O(1).$$

(4) $\{|t_0| \le t\}$, i.e., $y \ge \mu$ and it is fixed. Now, $e^{\varepsilon\left(t_0^2 - t^2\right)} \approx e^{-(y^2 - \mu^2)/\varepsilon} \to 0$. Then $\varepsilon\left(s^2 - t^2\right) \approx (s - t) \cdot 2y$ for s close to t, i.e., for those s, for which the contribution to the integral is dominant. We get $\int^t e^{2y(s-t)} ds \approx \frac{1}{2y}$. Hence,

$$r(t) \approx r_0 \left\{ r_0^2/2y \right\}^{-1/2} = \sqrt{y}.$$

We can summarize the above calculations.

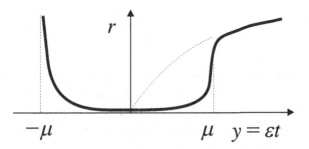

Fig. 4.12. Delay of loss of stability.

Theorem 4.25. *In case B, described by system (4.59) with $c < 0$, the phenomenon of delay in the loss of stability occurs. It relies upon the property that, as the variable y (which is the coefficient of stability of motion) changes from the negative value $y(t_0) = -\mu$ to the positive value μ, the system (with respect to z) is stable, and the change in the stability takes place for parameter $y = \mu$, whereby the amplitude of the oscillation increases as in the usual Andronov–Hopf bifurcation.*

The phenomenon of **delay in the loss of stability** can be explained physically. The variable y is negative for very long time, of order $1/\varepsilon$. Then the physical system will come very close to the equilibrium; close enough that it needs the same amount of time later to leave the equilibrium (see Fig. 4.12).

Remark 4.26. The most known example of a situation with relaxation oscillations, which arises for applications, is the FitzHugh–Nagumo system (2.28) from Chapter 2, when the parameters b, c, d are small.

4.4.4. *Duck solutions*

As we have explained above a typical evolution in a generic two-dimensional slow–fast system is like in Fig. 4.9 for the van der Pol system. Important property of those systems is that the spurt phenomenon is generic, i.e., in the system

$$\dot{x} = f(x, y) + O(\varepsilon), \quad \dot{y} = \varepsilon g(x, y) + O(\varepsilon^2),$$

we have

$$f''_{xx}(x_0, y_0)\, g(x_0, y_0) \neq 0 \quad \text{if} \quad f'_x(x_0, y_0) = 0.$$

But, when the functions f and g depend on some additional parameter (or parameters), the trajectories can different forms. French mathematicians (E. Benoit, J.-L. Callot, F. Diener, M. Diener) began to investigate families of slow–fast systems, using the so-called non-standard analysis.

An example of such family is as follows:

$$\dot{x} = y - x^2 - x^3, \quad \dot{y} = \varepsilon(a - y), \tag{4.61}$$

where a is a parameter and $\varepsilon > 0$ is our small parameter. Note that system (4.61) has singular point $p_a = (a, a^2 + a^3)$ with the linearization matrix

$$\begin{pmatrix} -a\,(2 + 3a) & 1 \\ -\varepsilon & 0 \end{pmatrix}$$

and the characteristic polynomial $P(\lambda) = \lambda^2 + a\,(2 + 3a)\,\lambda + \varepsilon$.

If $a \gg \varepsilon > 0$ then p_a is an attracting node. If $a = 0$ then p_a turns out to be an attracting focus. If $-\frac{2}{3} < a \ll -\varepsilon < 0$ then p_a is an unstable node, and if $a < -\frac{2}{3}$ then it is again a stable node.

Therefore, the scenario of bifurcations of a limit phase curve should look as follows. For $a > 0$ away form 0 we have a stable equilibrium which is still stable for $a = 0$. But as a becomes negative, with small $|a|$ then a slow–fast cycle appears near the point $(0, 0)$. It grows as a decreases, until it takes shape of the contour consisting of part of the graph of the function $y = x^2 + x^3$ for $-\frac{2}{3} \leq x \leq \frac{1}{3}$. Next, it takes the form like at Fig. 4.13. Finally, it takes the usual form like in Fig. 4.9.

The cycles like at Fig. 4.13 are called the **duck cycles** (from the French name "canard").

It should be underlined that the canard cycles take place in a very small region of the parameter space, a belongs to an interval of length about $e^{-1/\varepsilon}$.

The above scenario is typical for 1-parameter families of plane slow–fast systems and there exist a suitable theorem, which we do not formulate and refer the reader to [15].

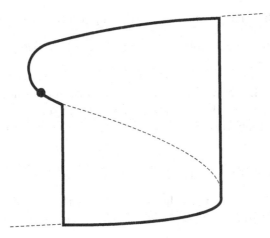

Fig. 4.13. Canard cycle.

4.5. Exercises

Exercise 4.27. Prove formulas (4.4).

Exercise 4.28. Solve the equation $\frac{dr}{d\theta} = \frac{r^2}{M}\sqrt{2\left(E - V(r)\right)}$ for $M > 0$ and $E_{\min} \leq E < 0$ and show that the orbits in the Kepler problem are ellipses.

Exercise 4.29. In the case of identical harmonic oscillators with $H = \frac{1}{2}|p|^2 + \frac{1}{2}|q|^2$, express the angular momentum via the actions I_1 and I_2 from Example 4.3(b).

Exercise 4.30. Show that the Euler–Poisson equations (4.8) are Hamiltonian with respect to the Poisson structure (4.9) with the Hamilton function (4.10). Show also the functions (4.11) have zero Poisson bracket with any variable M_i, Γ_j.

Exercise 4.31. Apply the Abelian integrals method (Section 4.1.3) to show that the van der Pol system $\dot{x} = y$, $\dot{y} = -x - a(x^2 - 1)y$ for small parameter $a > 0$ has exactly one limit cycle.

Exercise 4.32. Evaluate the integral $I_\omega(h) = \oint \omega$ from a degree n form over the circles $\{x^2 + y^2 = h\}$. Estimate the number of zeroes of I_ω.

Exercise 4.33. Prove formula (4.29) for the derivative of a Abelian integral.

Exercise 4.34. Prove the differential equations (4.30)–(4.31).

Exercise 4.35. Evaluate the integrals $I_j(h) = \oint x^j y \, dx$ along the compact oval $\gamma_h \subset \left\{ \frac{1}{2} y^2 + \frac{1}{3} x^3 - x = h \right\}$ at $h = 1$.

Exercise 4.36. Show that if g_s^t is a 2-parameter family of diffeomorphisms defining the evolution of a non-autonomous vector field $\dot{x} = v(t, x)$, which is periodic with a period T respect to time, then $g_{s+T}^{t+T} = g_s^t$.

Exercise 4.37. Prove formula (4.37).

Exercise 4.38. Show that there are exactly three collinear libration points in the three body problem.

Exercise 4.39. Consider a linear Hamiltonian system generated by the following homogeneous quadratic Hamiltonian:

$$H_2 = \frac{1}{2}(p_1 + q_2)^2 + \frac{1}{2}(p_2 - q_1)^2 - \frac{A}{2}q_1^2 - Bq_1 q_2 - \frac{C}{2}q_2^2.$$

(a) Prove that the characteristic polynomial of the corresponding matrix equals $P(\lambda) = \lambda^4 + a_2 \lambda^2 + a_4$, where

$$a_2 = 4 - A - C, \quad a_4 = AC - B^2.$$

(b) Show that the equilibrium point $q = p = 0$ is Lyapunov stable if: $a_2 > 0$, $a_4 > 0$ and $\Delta = a_2^2 - 4a_4 > 0$.

Exercise 4.40.

(a) In the case of a libration point $L_j = \left(q^{(j)}, p^{(j)} \right)$ apply the change $(q, p) \longmapsto \left(q_1 - q_1^{(j)}, q_2 - q_2^{(j)}, p_1 - q_2^{(j)}, p_2 + q_1^{(j)} \right)$ and show that you obtain a Hamiltonian system with the Hamilton function $H = \mathrm{const} + H_2 + \mathrm{h.o.t.}$, where H_2 is like in the previous problem.

(b) In the cases $L_{1,2,3}$ show that $A > 0$, $B = 0$ and $C < 0$ (because $f < 0$ in $\partial V / \partial q_2 = q_2 f$). Conclude that these libration points are unstable.

(c) In the case of the Lagrange points $L_{4,5}$ expand V in the Taylor series and show that $A = \frac{3}{4}$, $B = \zeta = 3\sqrt{3}\,(1 - 2\mu)\,/4$ and $C = \frac{9}{4}$; thus $P(\lambda) = \lambda^2 + \lambda^2 + \frac{27}{16} - \zeta^2$.

(d) Find the values of μ corresponding to the resonances 1:2 and 1:3.

Hint: We have $\omega_1^2 + \omega_2^2 = 1$ and $\omega_1^2 \omega_2^2 = 27/16 - \zeta^2$, where $|\omega_2| = n\,|\omega_1|$ for $n = 2, 3$.

Exercise 4.41. Show that the function H_0 from Eq. (4.38) is a Hamiltonian of a completely integrable system.

Exercise 4.42. Prove that the function (4.44) is the unique T–periodic solution to Eq. (4.43).

Exercise 4.43. Prove formula (4.60).

Chapter 5

Irregular Dynamics in Differential Equations

In the case of an autonomous vector field in $\mathbb{R}^2$, the phase portrait and the movement are fully determined; this is described in Section 2.5. But when the phase space is not so simple, interesting phenomena can occur.

For example, the constant vector field $\dot{\varphi}_1 = \omega_1$, $\dot{\varphi}_2 = \omega_2$ on the torus $\mathbb{T}^2 = \{(\varphi_1, \varphi_2)\}$ can have dense phase curves, i.e., when ω_2/ω_1 is irrational. Then the phase curves form a skein (like in Fig. 4.1 in Chapter 4) and the motion is quasi-periodic, which means that the solution returns roughly periodically to each small domain of the phase space. Moreover, you can reach any fixed small domain from any small domain.

In the Dynamic Systems Theory, such property is called the transitivity. The movement is not fully deterministic, because after a long time it is difficult to say where the evolving particle is. However, this is not a chaotic movement, because, if at the beginning we had a focused domain of the phase space, then this domain retains its focused shape during evolution. Meanwhile, in a chaotic movement, such a cell begins to "dissolve" in the phase space.

A good example of a situation illustrating the difference between transitivity and chaos are two glasses of water, such that one is infused with a small drop of oil and the other one with same amount of juice (Fig. 5.1). The oil droplet will drift, visit every spot in the water, and the juice will start to dissolve, filling the entire water domain evenly (this property is also called the mixing).

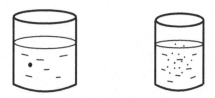

Fig. 5.1. Transitivity and mixing.

5.1. Translations on tori

We have encountered equations of the form

$$\dot{\varphi}_0 = \omega_0, \ldots, \dot{\varphi}_n = \omega_n, \tag{5.1}$$

where φ_j are coordinates on the torus $\mathbb{T}^{n+1}$ and ω_j are constant frequencies; we assume that $\varphi_j \in \mathbb{R}$ mod 1, i.e., the generating circles have length 1. For example, in linear systems in $\mathbb{R}^{2n+1}$ with imaginary eigenvalues or in invariant tori of completely integrable Hamiltonian systems.

The qualitative analysis of such system uses a corresponding return map $T : \mathbb{T}^n \longmapsto \mathbb{T}^n$, where $\mathbb{T}^n = \{\varphi_0 = 0\} \subset \mathbb{T}^{n+1}$ (provided $\omega_0 \neq 0$). This map is a translation,

$$T(\varphi) = \varphi + \alpha, \tag{5.2}$$

$\alpha = (\alpha_1, \ldots, \alpha_n)$; of course, we take $\varphi_j + \alpha_j$ mod 1. Important is the following property of the map (5.2):

The map T preserves the Lebesque measure $\mu = d^m\varphi$, which is probabilistic.

Therefore, we have a special case of the so-called **metric dynamical system**

$$(X, \mathfrak{M}, \mu, T), \tag{5.3}$$

where X is a set with a σ-algebra $\mathfrak{M}$ of measurable subsets, μ is a probabilistic measure on $\mathfrak{M}$ and $T : X \longmapsto X$ is a measurable mapping (endomorphism or automorphism if invertible) which preserves the measure: $T^{-1}(A) \in \mathfrak{M}$ and

$$\mu(T^{-1}(A)) = \mu(A),$$

if $A \in \mathfrak{M}$.

Definition 5.1. The dynamical system (5.3) is **ergodic** if any invariant subset $A \subset \mathfrak{M}$, i.e., $T^{-1}(A) = A$, is either X or is empty, i.e., $\mu(A) = 1$ or $\mu(A) = 0$.

There is a topological version of the latter notion.

Definition 5.2. If $T : X \longmapsto X$ is a homeomorphism of a metric space X then we say that it is **transitive** if for any two open subsets $A, B \subset X$ there exists $n \in \mathbb{Z}$ such that $T^n(A) \cap B \neq \varnothing$.

Theorem 5.3. *The automorphism* (5.2) *is ergodic if and only if the numbers* $1, \alpha_1, \ldots, \alpha_n$ *are non-resonant, i.e., there are no relations of the form* $k_0 + k_1\alpha_1 + \cdots + k_n\alpha_n$ *with* $(k_0, \ldots, k_n) \in \mathbb{Z}^{n+1} \backslash 0$. *In the latter case, this automorphism is also transitive.*

Proof. Assume that the numbers $1, \alpha_1, \ldots, \alpha_n$ are independent over $\mathbb{Q}$; note that this is equivalent to the independence of the frequencies ω_j in Eq. (5.1). It is enough to show that any bounded measurable function f on $\mathbb{T}^n$, invariant with respect to T, is constant a.e. Note that the indicator χ_A of an invariant subset is such function.

We expand this function into convergent Fourier series

$$f(\varphi) = \sum c_k e^{2\pi i(k, \varphi)},$$

where we sum over $k = (k_1, \ldots, k_n) \in \mathbb{Z}^n$. Its invariance, i.e., $f(\varphi + \alpha) = f(\varphi)$, leads to the relations

$$c_k = e^{2\pi i(k, \alpha)} c_k,$$

where $e^{2\pi i(k, \alpha)} \neq 1$ for $k \neq 0$ by the non-resonance assumption. So, all $c_k = 0$ for $k \neq 0$.

Assume now a relation $k_1\alpha_1 + \cdots + k_n\alpha_n = p$, with $p \in \mathbb{Z}$. Then the non-constant function $f(\varphi) = \exp(2\pi i(k, \varphi))$ is invariant.

To prove the topological transitivity in the non-resonant case, we note that for any open non-empty subset A of $\mathbb{T}^n$ the set $\bigcup_{m \in \mathbb{Z}} T^m(A)$ is invariant and, by its openness and the ergodicity of T, it is the whole $\mathbb{T}^n$. $\qquad \square$

We can explore further the above transitivity property. For a measurable subset $A \subset X$ and a point $x \in X$ denote $\tau(A, x, N)$ the number of points from the finite orbit $\{x, T(x), \ldots, T^{N-1}(\varphi)\}$ of an automorphism T which lie in A.

Definition 5.4. We say that the orbits of a metric automorphism T are **evenly distributed** if

$$\lim_{N \to \infty} \frac{1}{N} \tau(A, x, N) = \mu(A).$$

The above quantity $\frac{1}{N}\tau(A, x, N)$ can be interpreted as the following **ergodic mean** of a measurable function $f(x) = \chi_A(x)$:

$$S_N f(x) = \frac{1}{N} \sum_{j=0}^{N-1} f\left(T^j(x)\right). \tag{5.4}$$

The next result is a special case of a more general ergodic theorem (see Theorem 5.25).

Theorem 5.5. *In the case of non-resonant translation* (5.2), *we have*

$$\lim_{N \to \infty} S_N f(\varphi) = \int_{\mathbb{T}^n} f(\varphi) \, \mathrm{d}^n \varphi.$$

In particular, the orbits of this automorphism are evenly distributed.

Proof. Again, we expand f into the Fourier series. For the function $f_k(\varphi) = e^{2\pi i(k,\varphi)}$, we have

$$S_N f_k(\varphi) = \frac{1}{N} f_k(\varphi) \left\{ \sum_{j=0}^{N-1} \left(e^{2\pi i(k,\alpha)}\right)^j \right\} = \frac{1}{N} f_k(\varphi) \frac{1 - e^{2N\pi i(k,\alpha)}}{1 - e^{2\pi i(k,\alpha)}}$$

and it tends to zero as $N \to \infty$ when $k \neq 0$. $\qquad\qquad\square$

5.2. Diffeomorphisms of a circle

In the case of a vector field on a 2-torus close to a constant non-zero vector field, we can define the return map which is a diffeomorphism of a circle:

$$f : \mathbb{S}^1 \longmapsto \mathbb{S}^1,$$

where $\mathbb{S}^1 = \mathbb{R}^1/\mathbb{Z}^1$.[a] Of course, $f'(x) > 0$.

Define $F : \mathbb{R} \longmapsto \mathbb{R}$ the lift of f, i.e., such that the diagram

$$\begin{array}{ccc} \mathbb{R} & \overset{F}{\longmapsto} & \mathbb{R} \\ \downarrow \pi & & \downarrow \pi, \\ \mathbb{S} & \overset{f}{\longmapsto} & \mathbb{S} \end{array}$$

where $\pi(y) = y \mod 1$, is commutative. We can write

$$F(y) = y + \theta(y),$$

where the function $\theta(y)$ is periodic, $\theta(y+1) = \theta(y)$ and $\theta'(y) > -1$. Since the lift F is not uniquely defined, we can replace F with $F+1$, also the function θ is not uniquely defined, but we can ensure that

$$|\theta(y_1) - \theta(y_2)| < 2$$

for any y_1 and y_2. The same is true for any iteration $F^n(y) = y + \theta_n(y)$, $\theta_n(y) = \theta(y) + \theta(F(y)) + \cdots + \theta(F^{n-1}(y))$, which is the lift of $f^n(x)$.

Define the following ergodic mean:

$$S_N\theta(y) = \frac{1}{N}\sum_{j=0}^{N-1}\theta(F^j(y)) = \frac{1}{N}\theta_N(y).$$

Theorem 5.6. *The latter sequence is convergent to a number $\tau = \tau(F)$ which does not depend on y and its reduction modulo 1, called the **rotation number** of f, does not depend on the lift F.*

Proof. The independence on y is a consequence of the fact that $|\theta_N(y_1) - \theta_N(y_2)| < 2$. The further proof depends on whether f has periodic orbits or not.

If x_0 is a periodic point of f of period p and $x_0 = \pi(y_0)$, then x_0 is a fixed point of f^p and $F^p(y_0) = y_0 + q$, i.e., $\theta_p(y_0) = q$, for

---

[a]L. Siegel proved that for any vector field on $\mathbb{T}^2$ without singular points and closed orbits there exists a section with well-defined return map.

an integer q. Then $F^{Mp}(y_0) = y_0 + Mq$, or $\theta_{Mp}(y_0) = Mq$, and hence $\frac{1}{Mp}\theta_{Mp}(y_0) = \frac{q}{p}$. Next, for $N = Mp + r$, $0 < r < p$, we have $\theta_N(y_0) - \theta_{Mp}(y_0) = F^r\left(F^{Mp}(y_0)\right) - F^{Mp}(y_0) = \theta_r\left(F^{Mp}(y_0)\right)$ and is bounded independently on M and r; this easily implies that $\frac{1}{N}\theta_N(y_0) \to \frac{q}{p}$ as $N \to \infty$.

Assume now that there are no periodic points. For any $y \in \mathbb{R}$ and any integer $m > 0$, we have

$$r(m) < F^m(y) - y < r(m) + 1$$

for an integer $r(m)$. Moreover, these inequalities hold for any y, otherwise there would be a periodic point. Summing up these inequalities for $y = 0$, $y = F^m(0), \ldots, y = F^{(n-1)m}(0)$, we get

$$n \cdot r(m) < F^{nm}(0) - 0 < n \cdot (r(m) + 1).$$

Therefore, the two numbers $\frac{1}{m}F^m(0)$ and $\frac{1}{mn}F^{mn}(0)$ lie in the interval $\left(\frac{r(m)}{m}, \frac{r(m)+1}{m}\right)$, and hence $\left|\frac{1}{mn}F^{mn}(0) - \frac{1}{m}F^m(0)\right| < \frac{1}{m}$. By symmetry, we have the same inequality with m swapped with n. Therefore,

$$\left|\frac{1}{m}F^m(0) - \frac{1}{n}F^n(0)\right| < \frac{1}{m} + \frac{1}{n}$$

and the sequence $\left\{\frac{1}{n}F^n(0)\right\}$ is convergent. $\qquad\square$

If can be shown that the rotation number $\tau(f)$ does not change when we replace f conjugate diffeomorphism $h \circ f \circ h^{-1}$ for a diffeomorphism h (Exercise 5.47).

Proposition 5.7. *The rotation number $\tau(f)$ is irrational if and only if f does not have periodic points.*

Proof. It is enough to show that a diffeomorphism with rational rotation number has a periodic point. So, assume that $\tau(f) = \frac{q}{p}$ with $p > 0$.

Take the map

$$G(y) = F^p(y) - m,$$

such that $\lim_{N \to \infty} \frac{1}{N}G^N(y) = 0$. Thus, we reduce the problem to the case with zero rotation number.

Recall that the intersections of the graphic of G with the diagonal are projected to fixed points of $g = f^p$ in $\mathbb{S}^1$. If there are no such intersections then we have either $G(y) > y$ for all y, or $G(y) < y$ for all y. But then $\tau(g) > 0$ (respectively, $\tau(g) < 0$), because $\theta(y) > \varepsilon > 0$ (respectively, $\theta(y < -\varepsilon < 0$) for some $\varepsilon > 0$. □

The dynamics f in the case of rational rotation number is rather simple; it follows from the analysis of fixed points of $g = f^p$. Moreover, such diffeomorphism are typical, they form an open and dense subset of $C^1(\mathbb{S}^1, \mathbb{S}^1)$ (Exercise 5.49).

In the case of irrational rotation number τ, we have essentially two possibilities:

(i) f is topologically conjugated with the rotation

$$R_\tau(x) = x + \tau \mod 1;$$

(ii) there is a wandering interval.

In the first case, the dynamical system is transitive; every orbit of f is dense in $\mathbb{S}^1$ (Exercise 5.50).

Let us explain what we mean when we say about wandering interval in a simplest case.

Example 5.8. For the rotation R_τ by an irrational number τ consider an individual orbit:

$$\ldots, \; x_{-1} = R_{-\tau}(x_0), \; x_0, \; x_1 = R_\tau(x_0), \; x_2 = R_{2\tau}(x_0), \ldots.$$

All these points are pairwise different. Now, we replace each point x_j with an interval $I_j = [a_j, b_j]$ of positive length $|I_j| = b_j - a_j$; we must have $\sum_{j \in \mathbb{Z}} |I_j| \le 1$ and assume $\sum |I_j| = 1$. Define diffeomorphisms $\phi_j : I_j \longmapsto I_{j+1}$ which define a homeomorphism $\bigcup I_j \longmapsto \bigcup I_j$. We extend it to the whole circle by continuity.

The resulting homeomorphism $f : \mathbb{S}^1 \longmapsto \mathbb{S}^1$ has the rotation number $\tau(f) = \tau(R_\tau) = \tau$, but the points from the interior of I_0 have orbits which are not dense in $\mathbb{S}^1$.

Note that there can be more orbits of such wandering intervals.

Definition 5.9. The intervals I_j are called **wandering intervals**. The complement to the union of interiors of wandering intervals is denoted $\Omega(f)$ and consists of **non-wandering points**.

One can show that this non-wandering set equals to the ω-limit set and to the α-limit set of any point from $\mathbb{S}^1$ (see Definition 2.21 in Chapter 2).

Theorem 5.10. *Assume that $\tau = \tau(f)$ is irrational. Then there exists a continuous map $h : \mathbb{S}^1 \longmapsto \mathbb{S}^1$ which realizes a semi-conjugacy of f with the rotation R_τ in the sense that*

$$h \circ f = R_\tau \circ h.$$

It implies that the wandering intervals are sent by h to points and; if $\Omega(f) = \mathbb{S}^1$, then h is a homeomorphism.

Proof. We can assume that $0 \in \Omega(f)$. Let F be a lifting of F and assume that $\pi(0) = 0$. Then the inequalities

$$n_1\tau + m_1 < n_2\tau + m_2,$$

for $m_j, n_j \in \mathbb{Z}$, hold if and only if

$$F^{n_1}(0) + m_1 < F^{n_2}(0) + m_2.$$

Denote $y_n = F^n(0)$ and $x_n = \pi(y_n)$. We have two sets $A = \{y_n + m : m, n \in \mathbb{Z}\}$ and $B = \{n\tau + m : m, n \in \mathbb{Z}\}$ and a natural map $H : A \longmapsto B$ such that $H(y_n + m) = n\tau + m$. We prolong it to a map $H : \mathbb{R}^1 \longmapsto \mathbb{R}^1$ with the following properties:

$$H(y + 1) = H(y) + 1,$$
$$H(F(y)) = H(y) + \tau,$$
$$\operatorname{Im} H = \mathbb{R}^1,$$
$$H^{-1}(B) = A.$$

It sends the intervals J_k being components of $\pi^{-1}(I_k)$, I_k — a wandering interval, to points.

The map H is projected to a map h of the circle from the thesis of the theorem. $\square$

The fundamental result in this theory is the following result which guarantees the absence of wandering intervals.

Theorem 5.11 (Denjoy). *If f is of class C^2 and $\tau(f)$ is irrational, then f is transitive.*

Proof. Suppose that a sequence $\ldots, I_n, I_{n+1} = f(I_n), \ldots$ of wandering intervals of lengths $|I_n| > 0$ exists. Of course, $\sum |I_n| \leq 1$ and hence

$$|I_n|, |I_{-n}| \to 0 \quad \text{as } n \to \infty.$$

Of course,

$$|I_n| = |f^n(I_0)| = \int_{I_0} (f^n)'(x)\, dx,$$

where

$$(f^n)'(x) = f'(f^{n-1}(x)) f'(f^{n-2}(x)) \cdots f'(x)$$

$$= \exp \sum_{j=0}^{n-1} \phi(f^j(x)), \quad (n > 0),$$

$$(f^{-n})'(x) = 1/f'(f^{-n}(x)) \cdots f'(f^{-1}(x))$$

$$= \exp \left(-\sum_{j=1}^{n} \phi(f^{-j}(x))\right), \quad (n > 0),$$

$$\phi(x) = \ln f'(x).$$

In the case of a rotation R_τ by the irrational number τ, consider the orbit $\{\ldots, t_{-1} = R_\tau^{-1}(t_0), t_0 = 0, t_1 = R_\tau(t_0), \ldots\}$ of the point $t_0 = 0$ in $\mathbb{S}^1$. Let $N > 1$ be some moment of first return to a neighborhood of t_0; thus the points t_j, $|j| < N$, do not lie between t_0 and t_N (and also between t_0 and t_{-N}). (Such moments can be arbitrary large.) Assume that $0 < \theta := t_N - t_0 < \frac{1}{2}$; the opposite case is treated analogously.

Let $0 = t_0 = u_0 < u_1 < u_2 < \cdots < u_{N-1} < 1$ be the ordered truncated positive orbit of t_0, i.e., $\{t_0, t_1, \ldots, t_{N-1}\}$. We have $u_{j+1} - u_j > \theta$ and $1 - u_{N-1} > \theta$.

Consider now the truncated positive orbit of t_0 of length $2N$. It contains the points $v_0 = R_\tau^N(u_0) = u_0 + \theta$, $v_1 = u_1 + \theta, \ldots, v_{N-1} = R_\tau^{2N-1}(t_0) = u_{N-1} + \theta < 1$; in fact, we have $\{t_0, t_1, \ldots, t_{2N-1}\} = \{u_0, v_0, \ldots, u_{N-1}, v_{N-1}\}$ (by the definition of the first return moment). Therefore, we have the following ordering:

$$0 = u_0 < v_0 < u_1 < v_1 < \cdots < u_{N-1} < v_{N-1} < 1.$$

Denote now $x = R_\tau^N(t_0) = v_0$; then $\{v_0, \ldots, v_{N-1}\} = \{R_\tau^j(x) : j = 0, \ldots, N-1\}$ and $\{u_0, \ldots, u_{N-1}\} = \{R_\tau^{-j}(x) : j = 1, \ldots, N\}$ are truncated positive and negative orbits of x.

The same ordering takes place when we consider $t_j = f^j(t_0)$ with $x = f^N(t_0) = v_0 \in I_0$, $v_j \in \{t_0, \ldots, t_{N-1}\} = \{f^j(x) : j = 0, \ldots, N-1\}$ and $v_j \in \{t_N, \ldots, t_{2N-1}\} = \{f^{-j}(x) : j = 1, \ldots N\}$.

Consider the expression

$$\Phi_N(x) = \sum_{j=0}^{N-1} \phi\left(f^j(x)\right) - \sum_{j=1}^{N} \phi\left(f^{-j}(x)\right) = \sum_{j=0}^{N-1} \left(\phi(v_j) - \phi(u_j)\right),$$

where ϕ is defined above. Since f is of class C^2, and hence ϕ is of class C^1, we have

$$\left|\sum(\phi(v_j) - \phi(u_j))\right| \leq \sup|\varphi'| \cdot \sum|v_j - u_j| \leq C.$$

We claim that this gives a contradiction.

Indeed, on the one hand, the latter inequality implies that

$$A_N := \int_{I_0} \sqrt{(f^N)'} \cdot \sqrt{(f^{-N})'} dx = \int_{I_0} e^{\Phi(x)/2} dx \geq e^{-C/2} |I_0|.$$

On the other hand, by the Schwarz inequality, we have

$$A_N \leq \left(\int_{I_0} (f^N)' dx \cdot \int_{I_0} (f^{-N})' dx\right)^{1/2} = \sqrt{|I_N| \cdot |I_{-N}|},$$

which should tend to 0 as $N \to \infty$. $\qquad\qquad\qquad\qquad\qquad\square$

5.3. Intersections of invariant manifolds

Perhaps the simplest differential systems in which chaos can be observed are *periodic non-autonomous systems* of the form

$$\dot{x} = v(t, x), \quad x \in M, \quad v(t + T, x) = v(t, x), \qquad (5.5)$$

where M is a two-dimensional variety. As we know, such a system can be treated as autonomous in the expanded phase space $\mathbb{S}^1 \times M$.

Then it is convenient to work with the **monodromy map** (after the period)

$$\mathcal{P} : M \longmapsto M, \ \mathcal{P} = g_0^T,$$

where $g_s^t = \phi(t; x_0, s)$, $\phi(s; x_0, s) = x_0$, is a 2-parameter family of diffeomorphisms defining the evolution. In terms of the extended phase space, it is the return map to the hypersurface $\{0\} \times M$. In the monograph of J. Guckenheimer and P. Holmes [14], an example of the *Duffing system with external force*

$$\ddot{x} = x - x^3 + \varepsilon \left\{ \cos(\omega t) - ax \right\} \tag{5.6}$$

is analyzed. We will take a slightly different example.

Example 5.12 (Swing). It is the equation

$$\ddot{x} = -\sin x + \varepsilon \cos(\omega t), \tag{5.7}$$

where $\varepsilon \cos(\omega t)$ is a small periodic external force, with the period $T = 2\pi/\omega$. This can be interpreted as the equation of a swing with a girl, which performs periodic crouching (see Fig. 5.2). You can also treat this system as a subsystem of the four-dimensional autonomous system

$$\dot{x} = y, \ \ \dot{y} = -\sin x + \varepsilon z, \ \ \dot{z} = \omega u, \ \ \dot{u} = -\omega z.$$

However, let us focus on the extended phase space $\mathbb{S}^1 \times M$, where $M = \mathbb{S}^1 \times \mathbb{R}$ is a cylinder and we have

$$\dot{t} = 1, \ \dot{x} = y, \ \dot{y} = -\sin x + \varepsilon \cos(\omega t). \tag{5.8}$$

For a non-perturbed situation ($\varepsilon = 0$), the phase portrait is known (see Fig. 2.1 in Chapter 2); we present it in Fig. 5.3, where the upper

Fig. 5.2. Swing.

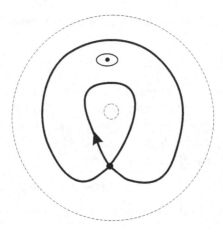

Fig. 5.3.   Phase portrait for pendulum.

Fig. 5.4.   Splitting of separatrices of saddle for a vector field.

and lower edges of the cylinder are shown as concentric dotted circles. We are interested in what will happen with the separatrix loop Γ of the saddle point $x = \pi$, $y = 0$ after switching on a perturbation.

If the perturbation were independent of time, then the expected phase portrait of the disturbed field would be like in Fig. 5.4, that is, the separatrices of the saddle point would be separated. However, in the case of a non-autonomous system, but periodic in time, the phase portrait of a non-perturbed system should be treated as the dynamics of the transformation of monodromy. In the perturbed system, the separatrices are not obliged to disconnect. We expect them to cross transversally, like in Fig. 5.5. We will show it in the following.

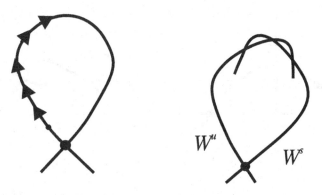

Fig. 5.5. Splitting of separatrices of saddle for a diffeomorphism.

The solution of the non-perturbed system, corresponding to the upper separatrix loop, is as follows:

$$x = x_0(t) = \pi - 4\tan^{-1}\mathrm{e}^{-(t-t_0)}, \quad y = y_0(t) = 2/\cosh(t - t_0) \quad (5.9)$$

(cf. Exercise 2.50 in Chapter 2). It has the property that $x(t_0) = 0$, $y(t_0) = 2$ and the value of the first integral

$$H(x, y) = y^2/2 - \cos x \qquad (5.10)$$

at those points is 1 (see Fig. 5.6).

For the study of the perturbed system ($\varepsilon \neq 0$) we will use the whole family of monodromy transformations

$$\mathcal{P}_z = g_z^{z+T} : M \longmapsto M, \quad z \in [0, T],$$

where $M = \mathbb{S}^1 \times \mathbb{R}$ is identified with a section $\{z\} \times M$ in the extended phase space $(\mathbb{R}/T\mathbb{Z}) \times M$. Each transformation $\mathcal{P}_z$ has its fixed point $q(z)$ (identified with $p(z) = q(z) + (2\pi, 0)$); this point depends on z and ε and lies close to the point $x = -\pi$, $y = 0$. Because it is a hyperbolic fixed point (saddle), it has its stable submanifold $W^s(p(z))$ and unstable $W^u(q(z))$ (see Fig. 5.6); of course, these submanifolds also depend on z and ε.

Choose the segment $S = \{x = 0, 1 < y < 3\}$ transversal to $W^s(p(z))$ and to $W^u(q(z))$. Let $\phi(t)$ (respectively, $\psi(t)$) be a solution with the initial condition $\phi(z) = S \cap W^s(p(z))$ (respectively $\psi(z) = S \cap W^u(q(z))$)). Of course, $\phi(t) \to p(z)$ at $t \to +\infty$ and $\psi(t) \to q(z)$ at $t \to -\infty$. Moreover, $\mathcal{P}_z(\phi(z)) = \phi(z + T)$ and $\mathcal{P}_z(\psi(z)) = \psi(z + T)$ (invariance of the submanifold).

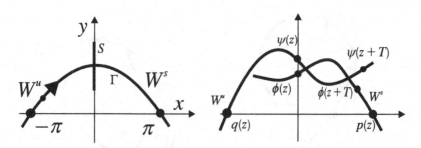

Fig. 5.6.   Derivation of Melnikov integral.

The intersection of the stable and unstable submanifolds is the situation when $\phi(z_0) = \psi(z_0)$ for corresponding z_0. As in the case of perturbations of autonomous Hamiltonian systems (see Section 4.1.4 in Chapter 4), the distance between $\phi(z)$ and $\psi(z)$ is calculated using the difference of the value of the first integral at these points,

$$\Delta H|_S = H(\psi(z)) - H(\phi(z)) = [H(\psi(z)) - H(q(z))]$$
$$+ [H(p(z)) - H(\phi(z))].$$

We have

$$H(\psi(z)) - H(q(z)) = \int_{-\infty}^{z} \dot{H} dt = \varepsilon \int_{-\infty}^{z} y \cos(\omega t)\, dt,$$

$$H(p(z)) - H(\phi(z)) = \int_{z}^{\infty} \dot{H} dt = \varepsilon \int_{z}^{\infty} y \cos(\omega t)\, dt.$$

Therefore, $\Delta H = \varepsilon \int_{-\infty}^{\infty} y \cos(\omega t)\, dt$, which we shall approximate by putting $y = y_0(t)$ from Eq. (5.9). We get

$$\Delta H = \varepsilon M(z) + O(\varepsilon^2) = \varepsilon \cdot 2 \int_{-\infty}^{\infty} \frac{\cos \omega t}{\cosh(t - z)} dt + O(\varepsilon^2), \quad (5.11)$$

where $M(z)$ is the so-called the **Melnikov function**.

It is not difficult to show the following lemma.

Lemma 5.13. *If $M(z_0) = 0$ and $M'(z_0) \neq 0$, then the submanifolds $W^s(p(z))$ and $W^u(q(z))$ intersect transversally at a point close to S.*

Proof. First, show that (as close to the phase curve of Eq. (5.9)) the submanifolds $W^s(p(z))$ and $W^u(q(z))$ lie horizontally in a neighborhood of the point $x = 0$, $y = 1$, i.e., are graphs of some functions of x. For $z = z_0$, we will treat them as graphs of functions F and G from a certain interval J (on the x-axis) to the segment S, where S is parametrized by $H|_S$.

Second, the transformations $\mathcal{P}_{z_0}$ and $\mathcal{P}_z$ are conjugated, $\mathcal{P}_z = g_{z_0}^z \circ \mathcal{P}_{z_0} \circ \left(g_{z_0}^z\right)^{-1}$. Conclude from this that $W^s(p(z)) = g_{z_0}^z(W^s(p(z_0)))$ and the same with W^u. The transformations $g_{z_0}^z$ are close to the transformations $g_0^{z-z_0}|_{\varepsilon=0}$ of the phase flow of the unperturbed system (5.8), which in the neighborhood of the point $x = 0$, $y = 2$ is roughly "to the right". Hence, it follows that, when z changes the submanifold $W^s(p(z))$ arises from the variety $W^s(p(z_0))$ by its "moving". It follows that, if $x_0(t)$ is given like in Eq. (5.9), then the function $F(x)$, whose graph is $W^s(p(z_0))$, can be given in a first approximation as

$$F(x) \approx H \circ \phi\left(x_0^{-1}(x)\right),$$

i.e., $F(x) \approx H(\phi(t))$ for t such that $x = x_0(t)$. Similarly, the graph of the function $G(x) \approx H \circ \psi\left(x_0^{-1}(x)\right)$ in first approximation defines $W^u(q(z_0))$. The difference $G(x) - F(x) \approx \Delta H \approx \varepsilon M(z)$. The transversality condition W^s and W^u results from the property: $\frac{d}{dx}(G - F) \neq 0$ for $G - F = 0$. $\qquad\square$

It turns out that the Melnikov integral from Eq. (5.11) is countable. Substituting $s = e^{z-t}$ (with $ds = -s\,dt$), we get

$$M(z) = -2 \int_0^\infty \frac{e^{i\omega z}s^{-i\omega} + e^{-i\omega z}s^{i\omega}}{1 + s^2}\,ds.$$

We calculate the integral $I = \int_0^\infty s^{i\alpha}(1 + s^2)^{-1}ds$ by the contour method. The integral along the contour from Fig. 5.7, with the radii of circles tending to 0 and ∞ respectively in the limit, is

$$(1 - e^{-2\pi i\alpha})I = 2\pi i\left\{\mathrm{res}_{s=i}s^{i\alpha}(1 + s^2)^{-1} + \mathrm{res}_{s=-i}s^{i\alpha}(1 + s^2)^{-1}\right\}$$

$$= \frac{2\pi i}{2i}\left(e^{-\pi\alpha/2} - e^{-3\pi\alpha/2}\right) = 2\pi e^{-\pi\alpha}\sinh(\pi\alpha/2).$$

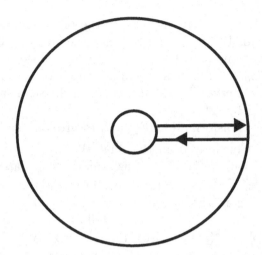

Fig. 5.7.   Integration contour.

That gives $I = \pi/(2\cosh(\pi\alpha/2))$ and

$$M(z) = 2\pi\frac{\cos(\omega z)}{\cosh(\pi\omega/2)}.$$

Of course, this function satisfies the requirement $M'|_{M=0} \neq 0$.

We have found at least one point r_0 of intersection of the stable and unstable manifolds of the fixed point $q = q(0)$ for the diffeomorphism

$$\mathcal{P} = \mathcal{P}_0 : U \longmapsto U,$$

where U is a certain neighborhood of the separatrix loop Γ of the saddle $x = \pm\pi$, $y = 0$, and $\mathcal{P}_0$ is a distinguished transformation of monodromy from the family $\{\mathcal{P}_z\}$ (with hyperbolic fixed points $q(z)$). But there are many more such points; they are of the form $r_n = \mathcal{P}^n(r_0)$, $n \in \mathbb{Z}$. As $n \to \infty$ and as $n \to -\infty$, the points r_n tend to the fixed point q_0.

However, the submanifolds $W^s = W^s(q(0))$ and $W^u = W^u(q(0))$ behave at least non-standardly. For example, the submanifold W^u passing through more and more further points r_n $(n \to \infty)$ begins to become more and more parallel to itself in vicinity of the saddle q (i.e., to the local unstable submanifold W^u_{loc}). Of course, between

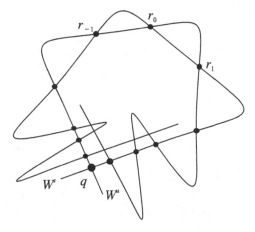

Fig. 5.8.   Intersection of stable and unstable submanifolds.

the points r_n and r_{n+1}, it makes a sharp turn. More or less the same happens with the submanifold W^s when passing through the points r_n for $n \to -\infty$ and between these points. In particular, the above pieces of W^u and W^s begin to intersect themselves at other (than r_n) points. One fears to imagine what is happening at further iterations; for example, pieces of W_u, parallel to W^u_{loc}, become longer and longer (see Fig. 5.8).

5.4.   Smale horseshoe and symbolic dynamics

Probably S. Smale was the first who understood the phenomenon from the end of the previous section and described it in rigorous mathematical terms. In Fig. 5.9, we see a (slightly curvilinear) "rectangle" R along the local stable submanifold W^s_{loc} which, under the action of sufficiently high iteration of the transformation $\mathcal{P}$, passes to a figure which cuts R in two places. One can choose the parameters defining the rectangle R so that this actually takes place (we do not do this, but we can refer the reader to books of R. Devaney [16], C. Robinson [17] and W. Szlenk [4]).

A model example of the transformation as in Fig. 5.9 is the transformation of Smale horseshoe shown in Fig. 5.10.

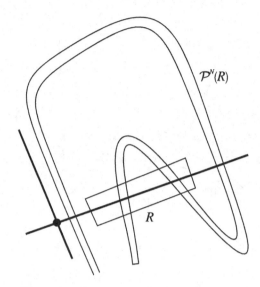

Fig. 5.9.   Generation of a horseshoe map.

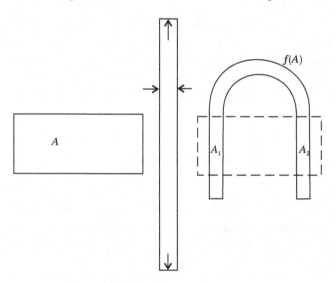

Fig. 5.10.   Smale horseshoe.

Definition 5.14 (Smale horseshoe). We have a (genuine) rectangle A on the plane with which we perform the following operations. First, we extend it in the vertical direction and squeeze it in the horizontal direction. Next, we bend our elongated rectangle and place it on the plane so that it intersects the initial rectangle A along two parallel vertical strips,

$$T(A) \cap A = A_1 \cup A_2.$$

In this way, we get a new figure, denoted $T(A)$, where $T : A \longmapsto T(A)$ is a **horseshoe diffeomorphism**.[b]

Smale horseshoe, although simply defined, is not so simple at all. It is easy to see that $T^2(A) \cap A$ consists of four vertical bars; more generally, $T^n(A) \cap A$ consists of 2^n horizontal strips (Exercise 5.51). On the other hand, $T^{-1}(A) \cap A = T^{-1}(A \cap T(A))$ consists of two horizontal strips; more generally, $T^{-m}(A) \cap A$, $n > 0$, consists of 2^m horizontal and thin strips (Exercise 5.52). Thus, $T^n(A) \cap T^{-m}(A)$, $m, n > 0$, consists of $2^n \times 2^m$ small rectangles.

Very important is the following set:

$$\Lambda = \bigcap_{n \in \mathbb{Z}} T^n(A). \tag{5.12}$$

It is easy to check that this set is invariant with respect to T : $T(\Lambda) = T^{-1}(\Lambda) = \Lambda$ (Exercise 5.53). We can say more about Λ and about $T|_\Lambda$, but first we should introduce one definition.

Definition 5.15. Let $\Sigma = \Sigma_k = \{1, \ldots, k\}^{\mathbb{Z}}$ be a countable Cartesian product of a fixed k-elements set; it consists of strings (sequences) $a = (\ldots, a_{-1}, a_0, a_1, \ldots)$, $a_j \in \{1, \ldots, k\}$. We define the transformation $\sigma : \Sigma \longmapsto \Sigma$ as follows:

$$\sigma(a)_j = a_{j+1}.$$

The dynamic system (Σ, σ) defined above is called the **symbolic system** (or the **shift map**).

---

[b]You can extend this transformation. Let us attach half-discs to the bottom and top bases of A and denote the new figure by M. Let us extend f onto the half-discs, so that their images adhere to the lower ends $f(A)$. Assuming that the new figure lies completely in M, we get a well-defined diffeomorphism $f : M \longmapsto M$.

On the space Σ a product topology is introduced, such neighborhoods of the given sequence $a = (\ldots, a_{-1}, a_0, a_1, \ldots)$ are the so-called **cylindrical sets** of the form

$$C_{a_{-M}, \ldots, a_N} = \{b = (\ldots, b_{-1}, b_0, b_1, \ldots) : b_i = a_i, \ i \in \{-M, \ldots, N\}\}$$

(for fixed M, N). Σ is also a metric space, because the distance between two strings is $\mathrm{dist}\,(a, b) = \sum_{n \in Z} 2^{-|n|} |a_n - b_n|$.

One has the following theorem.

Theorem 5.16. *There is a homeomorphism* $\Phi : \Lambda \longmapsto \Sigma_2$, *which conjugates* σ *with* $T|_\Lambda$:

$$\sigma \circ \Phi = \Phi \circ T.$$

Proof. The transformation Φ is easy to define. If $x \in \Lambda$, we put $\Phi(x) = (\ldots, a_{-1}, \ a_0, a_1, \ldots)$, where

$$a_n = 1 \text{ if } T^n(x) \in A_1 \quad \text{and} \quad a_n = 2 \quad \text{if } T^n(x) \in A_2.$$

The conjugation property is checked directly (Exercise 5.55). Therefore, it remains only to show the continuity and the invertibility of the transformation Φ.

These two properties result from hyperbolicity of the horseshoe transformation: in the horizontal direction we have contraction with a ratio $\lambda_1 < 1$ and in the vertical direction we have expansion with a ratio $\lambda_2 > 1$. Thus, the rectangles appearing at the location of points x, i.e.,

$$\left\{x : T^{-M}(x) \in A_{a_{-M}}, \ldots, T^N(x) \in A_{a_N}\right\} \tag{5.13}$$

become exponentially small with M and N very large. We get only one point in the limit (invertibility). Small sizes of the sets (5.13) correspond to the small cylindrical sets in Σ_2; this is exactly the continuity of Φ and Φ^{-1}. $\qquad\square$

Since Λ is the only invariant set in the rectangle A, all interesting dynamics of the transformation of the horseshoe is limited to the dynamics of $T|_\Lambda$. Due to the above theorem, it is the same dynamics as for the symbolic transformation of σ on Σ. On the other hand, the symbolic transformation is pleasant to study. It has the following interesting properties.

Proposition 5.17. *The periodic points of* σ *are dense in the symbolic space* Σ.

Proof. Let $a = (\ldots, a_0, a_1, \ldots) \in \Sigma$. For a large $N > 0$ all sequences $b = (\ldots, b_{-1}, b_0, b_1, \ldots)$ such that $b_{-N} = a_{-N}, \ldots, b_N = a_N$ are close to a. Thus, the sequence formed from the block $(a_{-N}, \ldots, a_N)$ (of length $2N + 1$) and periodically repeated is also close to a. It corresponds to a periodic point of σ of period $2N + 1$. $\qquad\qquad\square$

Proposition 5.18. *The dynamical system* (Σ, σ) *is transitive, i.e., for any open subsets* $U, V \subset \Sigma$ *there exists* $n > 0$ *such that* $T^n(U) \cap V \neq \varnothing$.

Proof. It suffices to consider the case when U and V are cylindrical sets defined by means of blocks $(a_1, \ldots, a_M)$ and $(b_1, \ldots, b_N)$. Then just take any sequence with the block $(a_1, \ldots, a_M, b_1, \ldots, b_N)$ (of length $M + N$). $\qquad\qquad\square$

5.5. Ergodic aspects of dynamical systems

5.5.1. *Invariant measures*

Since orbits of some high-dimensional dynamical systems behave irregularly, a natural approach is to study their statistics. This is realized using corresponding measures which should be invariant and probabilistic.

Recall that a probabilistic measure μ on X is invariant for a transformation (endomorphism) T of X if $\mu\left(T^{-1}(A)\right) = \mu(A)$; i.e., $T_*\mu = \mu$.

In the case of an autonomous differential system, i.e., a vector field v on a differentiable manifold, a Borel probability measure μ is invariant if it is invariant with respect to the flow $\left\{g_v^t\right\}$: $\left(g_v^t\right)_* \mu = \mu$.

Example 5.19. For a Hamiltonian system $\dot{q} = H'_p$, $\dot{p} = -H'_q$, $(q, p) \in \mathbb{R}^{2n}$, the Lebesque measure $\mathrm{d}^n p \mathrm{d}^n q$ is invariant, because the divergence of this vector field is zero (see Theorem 6.24 in the next chapter).

More generally, the same Theorem 6.24 says that, if we have a vector field v on $\mathbb{R}^m$ and a function $\rho \geq 0$ such that $\mathrm{div}(\rho v) \equiv 0$, then the measure $\rho \mathrm{d}^m x$ is invariant.

In these examples, the measure is usually infinite. But often we can generate from it a probabilistic measure. For example, if some level hypersurface $M_h = \{H = h\}$ is compact, then the so-called Gelfand–Leray form

$$\eta = dp \wedge dq/dH,$$

which is a $(2n - 1)$-form such that $dH \wedge \eta = dp \wedge dq$ and which is uniquely defined on M_h, defines a finite invariant measure μ_h on M_h, which can be normalized to an invariant probability measure. Sometimes, e.g., in the statistical mechanics, one deals with invariant probability measures of the form

$$\mu = \rho(h) \, dh d\mu_h$$

with a density $\rho(h)$.

Also, in the case of a completely integrable Hamiltonian system, on an invariant torus one has the invariant measure

$$dp \wedge dq/dF_1 \wedge \cdots \wedge dF_n,$$

where F_j are the first integrals in involution.

We have the following general result in this subject.

Theorem 5.20 (Bogolyubov–Krylov). *Let X be a compact metric space and $T : X \longmapsto X$ be a continuous mapping. Then there exists an invariant Borel measure for T.*

Proof. We have the bounded linear operator

$$T^* : C(X) \longmapsto C(X),$$

$(T^*\varphi)(x) = \varphi(T(x))$. Above $C(X)$ is the Banach space of continuous functions on X with the norm $\|\varphi\| = \sup |\varphi(x)|$. The conjugate to $C(X)$ is the space $M(X) = C^*(X)$ of charges on X (measures which can take negative values), and the dual to T^* is the operator T_*, also continuous.

The space $M(X)$ contains a subset $P(X)$ of probabilistic measures. The operator T_* preserves this set,

$$T_* : P(X) \longmapsto P(X). \tag{5.14}$$

Recall also that $M(X)$ is equipped with a weak topology, with respect to which T_* is continuous. In this topology, the set $P(X)$ is a closed subset of the infinite product $\prod_{\varphi:\|\varphi\|=1} I_\varphi$, $I_\varphi = [\inf \varphi, \sup \varphi]$, with the Tikhonov product topology. Therefore, $P(X)$ is compact. Of course, $P(X)$ is also convex.

Therefore, by a corresponding theorem from the functional analysis, the Tikhonov fixed point theorem, the operator (5.14) has a fixed point μ_T.

The measure μ_T, which is quite abstract, can be obtained from the following sequence of measures. Take an arbitrary probabilistic Borel measure ν on X and define the averages $\nu_N = \frac{1}{N} \sum_{j=0}^{N-1} T_*^j \nu$. By the compactness of $P(X)$, there exists a convergent subsequence $\{\nu_{N_k}\}$. □

Example 5.21. For the space $X = \Sigma_k = \{1, \ldots, k\}^{\mathbb{Z}}$ define the following **Bernoulli measure** μ such that on the cylindrical subsets $C_{b_{-M}, \ldots, b_N}$ it equals $(1/k)^{N+M+1}$. It is clear that it is invariant with respect to the shift σ.

This example can be generalized as follows. Consider the set of infinite sequences

$$X = \Sigma_k^+ = \{1, \ldots, k\}^{\mathbb{Z}_+} = \{a = (a_0, a_1, \ldots)\}$$

and the collection of *transition probabilities* $0 \leq p_{i,j}$, $i, j = 1, \ldots, k$, such that

$$\sum_j p_{i,j} = 1. \tag{5.15}$$

We have the shift map $\sigma((a_0, a_1, \ldots)) = (a_1, a_2, \ldots)$. Finally, for a given *initial distribution* $\{p_1, \ldots, p_k\}$, $0 \leq p_j$, $\sum p_j = 1$, we define a probabilistic measure on the cylindrical subsets by the formula

$$\mu(C_{b_0, \ldots, b_N}) = p_{b_0} p_{b_0, b_1} \cdots p_{b_{N-1}, b_N}. \tag{5.16}$$

In particular, we have $\mu(\{a_1 = j\}) = \sum_i p_i p_{i,j}$.

Definition 5.22. The above dynamical system $(\Sigma_k^+, \mathcal{C}, \sigma, \pi)$, where $\mathcal{C}$ is the σ-algebra generated by the cylindrical subsets, is called the (one-sided) **Markov chain**.

Such Markov chains appear as invariant subsets of dynamical systems more complex than the Smale's horseshoe.

Theorem 5.23. *For a given system $\{p_{j,j}\}$ of transition probabilities, there exists an initial distribution $\{p_j\}$ such that*

$$p_j = \sum_i p_i p_{i,j},$$

which means that the above measure μ is σ-invariant and the system $\left(\Sigma_k^+, C, \sigma, \mu\right)$ is a genuine dynamical system (called the stationary Markov chain).

Proof. We define the following transfer operator $\mathcal{P} : \mathbb{R}^k \longmapsto \mathbb{R}^k$:

$$(\mathcal{P}p)_j = \sum_i p_{i,j} p_i.$$

It is continuous, positive (if all $p_j \geq 0$ then also all $(\mathcal{P}p)_j \geq 0$) and preserves the property $\sum p_i = 0$.

By the Brouwer fixed point theorem it has a fixed point p^*.

Finally, we note that, like in the final remark in the proof of Theorem 5.20, we can consider the ergodic averages

$$S_N p = \frac{1}{N} \sum_{j=0}^{N-1} \mathcal{P}^j p.$$

Assume also that all $p_{i,j} > 0$. It turns out that these averages tend to p^* (without taking a subsequence) and this result is known as the *ergodic theorem for Markov chains*. $\qquad\square$

5.5.2. Recurrence and ergodic theorem

We begin with the following classical theorem.

Theorem 5.24 (Poincaré recurrence). *Let $T : X \longmapsto X$ be a measurable map that preserves a probabilistic measure μ. Then for any measurable subset A with $\mu(A) > 0$ for almost any point $x \in A$ its positive orbit visits A, i.e., $T^n(x) \in A$ for some $n > 0$.*

Proof. Let the subset $B \subset A$ consists of those points in A which never return to A. In particular, for any $n > 0$ and $x \in B$, we have $T^n(x) \notin B$, or $x \notin T^{-1}(B)$; so, $T^{-n}(B) \cap B = \varnothing$. Then $T^{-k}(B) \cap T^{-k-n}(B) = t^{-k}(B \cap T^{-n}(B)) = \varnothing$ for any $k > 0$. Therefore, the sets $T^{-n}(B)$ are pairwise disjoint.

But they have the same measure. If B would have positive measure, then this would contradict the finiteness of the measure μ. □

The above recurrence property is sometimes mistaken with the transitivity property and is associated with some apparent paradoxes.

For example, a circle diffeomorphism with rational rotation number and hyperbolic periodic points is not transitive. However, any invariant measure is supported on periodic points and is discrete.

In another example, one considers two chambers of the same volume with gas. At the initial moment $t = 0$ all particles of the gas are localized in the left chamber while the second chamber is empty. Next, we open the left chamber and the gas quickly fills the both chambers and never returns completely to the left chamber.

The latter paradox is explained by the fact that the average time of return to the subset A from the recurrence theorem is equal $\frac{1}{\mu(A)}$. In the latter example, $\mu(A)$ is of order $(1/2)^N$, where N is large, like the Avogadro number $\sim 6 \cdot 10^{23}$.

The next theorem is a generalization of Theorem 5.5 about equality of averages for a transitive toral automorphism. We present it without proof which is rather technical and not intuitive; we refer the reader to the books [4,17,18].

Theorem 5.25 (Birkhoff ergodic). *Let* $(X, \mathfrak{M}, T, \mu)$ *be a dynamical system. Then for any function* $f \in L^1(X, \mathfrak{M}, \mu)$ *the sequence* $\{S_N f\}$ *of ergodic averages* (5.4) *is convergent a.e. to a function* $\bar{f} \in L^1(X, \mathfrak{M}, \mu)$ *such that*

$$T^* \bar{f} = \bar{f} \ a.e.,$$

$$\int_X \bar{f} \mathrm{d}\mu = \int_X f \mathrm{d}\mu.$$

In particular, if the dynamical system is ergodic, then $\bar{f}(x) \equiv \int_X f \mathrm{d}\mu \ a.e.$

Remark 5.26. The limit function $\bar{f}$ has the following probabilistic interpretation. Consider the σ-subalgebra $\mathfrak{F}$ of $\mathfrak{M}$ consisting of measurable subsets A which are T-invariant, $T^{-1}(A) = A$. Then $\bar{f}$ is the conditional expectation with respect to the algebra $F : \bar{f} = \mathbb{E}(f|\mathfrak{F})$.

For example, if T has a first integral H such that $T^*H = H$, and the restriction of T to the levels $X_h = \{H(x = h)\}$ is "metrically" transitive, then $\mathfrak{F}$ is generated by sets of the form $\{H(x) \in C\}$, where C are Borel subsets of $\mathbb{R}$. In this case, $\bar{f}(x) = \tilde{f}(H(x))$ for a measurable function $\tilde{f}$ on $\mathbb{R}$.

5.5.3. *Mixing and entropy*

The following property of a dynamical system is stronger than the ergodicity or the transitivity.

Definition 5.27. A dynamical system (X, M, μ, T) is **mixing** if, for any measurable subsets A and B of positive measure, we have

$$\lim_{n \to \infty} \frac{1}{\mu(B)} \mu\left(T^{-n}(A) \cap B\right) = \mu(A).$$

The next quantity measuring the chaotic behavior of a metric dynamical system is its entropy defined as follows.

Let $\xi = \{A_1, \ldots, A_n\}$ be a *partition* of X with the following properties:

$$A_i \cap A_j = \varnothing \text{ for } i \neq j, \quad \bigcup_i A_i = X.$$

If $\xi = \{A_i\}$ and $\eta = \{B_1, \ldots, B_m\}$ are two partitions, then their joint is the partition

$$\xi \vee \eta = \{A_i \cap B_j : i = 1, \ldots, n, j = 1, \ldots, m\}.$$

We say that a partition ξ is finer than a partition η, $\xi \preceq \eta$, if any element of η is a union of elements from ξ, $B_j = \bigcup_{i \in I_j} A_i$, $I_j \subset \{1, \ldots, n\}$. Thus, $\xi \vee \eta \preceq \xi, \eta$.

For a partition $\xi = \{A_i\}$ we define its entropy as

$$H(\xi) = \sum \mu(A_i) \ln(1/\mu(A_i)). \tag{5.17}$$

Lemma 5.28. *We have* $H(\xi \vee \eta) \leq H(\xi) + H(\eta)$.

Proof. We use the conditional measures $\mu(A_i|B_j) = \mu(A_i \cap B_j)/\mu(B_j)$ and the concavity of the function $\Phi(x) = -x \ln x$.
We have

$$H(\xi \vee \eta) = -\sum \mu(A_i \cap B_j) \ln(A_i \cap B_j)$$

$$= -\sum_i \left[\sum_j \mu(B_j) \cdot \mu(A_i|B_j)(\ln \mu(B_j) + \ln \mu(A_i|B_j)) \right]$$

$$= H(\eta) + \sum_i \sum_j \mu(B_j) \Phi(\mu(A_i|B_j))$$

$$\leq -H(\eta) + \sum_i \Phi(\mu(A_i))$$

$$= H(\eta) + H(\xi). \qquad \square$$

Assume now that a measurable map T acts on X. With a partition $\xi = \{A_i\}$ we associate the partitions $T^{-k}\xi = \{T^{-k}(A_i)\}$ and $\xi \vee T^{-1}\xi \vee \cdots \vee T^{-n}\xi$.

Proposition 5.29. *There exists*

$$\lim_{n \to \infty} \frac{1}{n} H\left(\xi \vee T^{-1}\xi \vee \cdots \vee T^{-n}\xi\right) = h(T,\xi).$$

Proof. By Lemma 5.28, the numbers $a_n = H\left(\xi \vee T^{-1}\xi \vee \cdots \vee T^{-n}\xi\right)$ satisfy the relations $a_{m+n} \leq a_m + a_n$. Then it is an exercise from the Calculus (Exercise 5.56). $\qquad \square$

Definition 5.30. The **metric entropy** of a dynamical system $(X, \mathfrak{M}, \mu, T)$ is

$$h(T) = h_\mu(T) = \sup_\xi h(T,\xi).$$

Example 5.31. Consider the symbolic system $(\Sigma_k, \mathcal{C}, \mu, \sigma)$, where μ is the Bernoulli measure from Example 5.21.
Consider two cylindrical sets A and B defined by fixing values of the sequences $a = (\ldots, a_0, \ldots)$ for indices from the sets $I_A \subset \mathbb{Z}$ and $I_B \subset \mathbb{Z}$. If $n > 0$ is sufficiently large, then the sets $I_A + n$ and I_B become disjoint. Therefore, the set $\sigma^{-n}(A) \cap B$ is defined

by the indices' set $(I_A + n) \cup I_B$ of cardinality $|I_A| + |I_B|$. We have $\mu(A) = (1/k)^{|I_A|}$, $\mu(B) = (1/k)^{|I_B|}$ and

$$\mu\left(\sigma^{-n}(A) \cap B\right) = (1/k)^{|(I_A+n)\cup B|} = \mu(A) \cdot \mu(B).$$

This means that the Bernoulli shift, as well as the horseshoe map on the invariant set Λ with the measure induced from the Bernoulli measure on Σ_2, are mixing.

Let $\xi = \{A_1, \ldots, A_k\}$ be the partition by the sets $A_i = \{a : a_0 = i\}$. Of course, $H(\xi) = \ln k$. Also $H(\xi \vee T^{-1}\xi \vee \cdots \vee T^{-n}\xi) = \ln k$.

Moreover, it can be shown that the entropy $h(T)$ from Definition 5.30 equals $h(T, \xi)$, provided ξ is a generating partition in the sense that the partitions $T^{-m}\xi \vee \xi \vee \cdots \vee T^n\xi$ (for any $m, n > 0$) generate the whole σ–algebra $\mathfrak{M}$.

The above partition generates all cylindrical sets. Therefore,

$$h(\sigma) = \ln k.$$

Example 5.32. The translations on tori are not mixing and have zero entropy (Exercise 5.57).

We finish this section with some general comments.

The notion of metric entropy defined above is an analogue of an entropy from the Statistical Mechanics. That entropy measures the number of degree of freedom of a system of large number of particles.

In the Statistical Mechanics, there exists a notion of the partition function of such large system. The logarithm of the partition function divided by the number of particles is known as the free energy.

D. Ruelle was the first who adapted the thermodynamic formalism from the Statistical Mechanics to the Dynamical Systems Theory. The dynamical version of the free energy is defined similarly like the above entropy, but using averaging of some function (Gibbs density) and the obtained analogue of the entropy is called the pressure.

Next, there exist a notion of a topological entropy $h_{\text{top}}(T)$ of a continuous map $T : X \longmapsto X$ of a metric space; it uses coverings by open subsets, analogues of partitions. It turns out that

$$h_{\text{top}}(T) = \sup h_\mu(T),$$

where the supremum is taken over all invariant Borel measures for T. We refer the reader to the books [4,17] for more information on this subject.

5.6. Smooth dynamical systems

5.6.1. *Hyperbolicity*

The subset $\Lambda \subset \mathbb{R}^2$, invariant for the Smale horseshoe transformation, has one more important property. Namely, it is hyperbolic, which means that induced linear transformations $DT(x) : T_x\mathbb{R}^2 \longmapsto T_{f(x)}\mathbb{R}^2$ are hyperbolic (they have one eigenvalue $\lambda_1 \in (0,1)$ and a second $\lambda_2 > 1$). This is generalized as follows.

Let $T : M \longmapsto M$ be a diffeomorphism of a manifold M and let $\Lambda \subset M$ be a closed invariant subset.

Definition 5.33. The subset Λ is **hyperbolic** if the restriction of the tangent bundle TM to Λ is split into a Whitney sum

$$T_\Lambda M = E^u \oplus E^s$$

into subbundles which are invariant with respect to DT, $DT(x) :$ $E_x^{u,s} \longmapsto E_{T(x)}^{u,s}$, and there exist constants $C > 0$ and $0 < \lambda < 1$ such that

$$\|DT^n(x)w\| \geq C\lambda^{-n}, \ (w \in E_x^u), \ \|DT^n(x)w\| \leq C\lambda^n,$$

$$(w \in E_x^s) \tag{5.18}$$

for $n \geq 0$.

In the case of a flow $\{g_v^t\}$ generated by a vector field $v(x)$ on M, an invariant subset Λ is hyperbolic if

$$T_\Lambda M = E^u \oplus E^s \oplus E^0,$$

where E^u and E^s are invariant subbundles satisfying bounds analogous to (5.18) and the third bundle is tangent to the phase curves of v, $E_x^0 = \mathbb{R}v(x)$.

The next result is a generalization of the Hadamard–Perron theorem from Chapter 1, with a similar proof which we skip. Analogous result holds in the case of an invariant hyperbolic set for a flow.

Theorem 5.34. *If Λ is a hyperbolic invariant subset for a diffeomorphism T, then for any $x \in \Lambda$ there exist invariant immersed submanifolds $W^u(x)$, $W^s(x) \subset M$, such that*

$$W^u(x) = \{y : \operatorname{dist}(T^n(x), T^n(y)) \to 0 \ as \ n \to -\infty\},$$

$$W^s(x) = \{y : \operatorname{dist}(T^n(x), T^n(y)) \to 0 \ as \ n \to \infty\}.$$

Definition 5.35. A **Markov partition** of an invariant hyperbolic set Λ for a diffeomorphism T is a finite collection $\xi = \{A_1, \ldots, A_n\}$ of subsets $A_j \subset \Lambda$ with the following properties:

- $\Lambda = \bigcup A_j$;
- $A_j = \mathrm{cl}\,(\mathrm{int} A_j)$ and $A_i \cap A_j = \partial A_i \cap \partial A_j$ if $i \neq j$;
- the intersections $W_{\mathrm{loc}}^{u,s}(x) \cap A_j$ of local invariant manifolds behave correctly with respect to T, if $x \in \mathrm{int} A_i \cap T^{-1}(\mathrm{int} A_j)$ then

$$A_j \cap W_{\mathrm{loc}}^u(T(x)) \subset T(A_i \cap W_{\mathrm{loc}}^u(x)),$$
$$T(A_i \cap W_{\mathrm{loc}}^s(x)) \subset A_j \cap W_{\mathrm{loc}}^s(T(x)).$$

Above int, cl and ∂ denote the interior, the closure and the boundary of a subset of Λ (with a topology in Λ, not in M). We can imagine the sets A_j as rectangles with sides along the local invariant manifolds. Note also that the "global" invariant manifolds from Theorem 5.34 can be dense in M, like in the case of Anosov diffeomorphisms considered in the next section.

In the case of Smale horseshoe, the corresponding Markov partition is $\{A_1 \cap \Lambda, A_2 \cap \Lambda\}$, where A_1 and A_2 are the rectangles from Fig. 5.9.

The existence of a Markov partition allows to construct a conjugacy of $T|_\Lambda$ with a suitable Markov chain.

The following result is crucial in this subject, but is not easy; so, we state it without a proof.

Theorem 5.36 (Bowen). *For a hyperbolic invariant subset, there exist a Markov partition.*

Recall the following notion from Definition 5.9.

Definition 5.37. For a diffeomorphism T of a manifold M the wandering points are such that some their neighborhoods U have disjoint orbit, $T^m(U) \cap T^n(U) = \varnothing$ for $m \neq n$. The set $\Omega(T)$ consisting of non-wandering points is called the **non-wandering set** for T.

Definition 5.38. A diffeomorphism T of a manifold M satisfies **Axiom A** if (i) the set $\Omega(T)$ is hyperbolic and (ii) the periodic points of T are dense in $\Omega(T)$.

A flow $\{g_v^t\}$ satisfies **Axiom A** if (i) the equilibrium points are hyperbolic, (ii) $\Omega\left(g_v^t\right)$ is a union of the set of equilibrium points and a closure Λ of the set of closed phase curves and (iii) Λ is hyperbolic.

Remark 5.39. The latter definition is a generalization of the notion of **Morse–Smale diffeomorphisms**, which satisfy the following conditions: (i) $\Omega\left(T\right)$ is finite (and consists of periodic points), (ii) the periodic points are hyperbolic and (iii) the stable and unstable invariant manifolds of periodic points intersect one another transversally.

We state, without proof, the following result (cf. Theorem 5.42).

Theorem 5.40 (Robbin). *If a diffeomorphism T of M satisfies Axiom A and is such that for any $x, y \in \Omega\left(T\right)$ the invariant manifolds $W^u\left(x\right)$ and $W^s\left(y\right)$ intersect transversally, then T is structurally stable.*

5.6.2. *Anosov diffeomorphisms*

The invariant hyperbolic set Λ for the Smale horseshoe diffeomorphism is very thin (its Hausdorff dimension depends on λ_1 and λ_2) and it is certainly not a manifold (even locally). But there exist chaotic dynamic systems with a hyperbolic structure on the whole manifold.

They are the so-called **Anosov diffeomorphisms** and **Anosov flows**; the conditions from Definition 5.33 hold on the whole $=\Omega\left(T\right)$.[c]

Example 5.41 (Hyperbolic torus automorphism). Let us identify the two-dimensional torus with the plane quotiented by the integer lattice, $\mathbb{T}^2 = \mathbb{R}^2/\mathbb{Z}^2$. The matrix

$$A = \begin{pmatrix} 2 & 1 \\ 1 & 1 \end{pmatrix} \tag{5.19}$$

defines a transformation of the plane and has the property that points with integer coordinates are sent into similar points. Thus, it defines

---

[c]In the Russian literature, they are called the U-systems.

a transformation $T : \mathbb{T}^2 \longmapsto \mathbb{T}^2$. Since the determinant of our matrix is equal to 1, the inverse transformation preserves the lattice; thus T is a diffeomorphism.

The transformation T has exactly one fixed point, corresponding to the point $(0, 0)$. However, equations for periodic points with period 2 take the form $4x_1 + 3x_2 = m_1$, $3x_1 + x_2 = m_2$, $m_{1,2} \in \mathbb{Z}$. It is not difficult to see that this gives 25 solutions. In general, with growing n the number of periodic points for T with period $\leq n$ grows to infinity; in particular, points with rational two coordinates are periodic (Exercise 5.58).

The derivative matrix $DT(x) : T_x \mathbb{T}^2 \longmapsto T_{T(x)} \mathbb{T}^2$ at each point x is the same and equal to A. On the other hand, the matrix A is hyperbolic, with eigenvalues $\lambda_1 = \frac{1}{2}(3 - \sqrt{5}) < 1$ and $\lambda_2 = \frac{1}{2}(3 + \sqrt{5}) > 1$. Thus, T has a (uniform) hyperbolic structure. (This property enters into the definition of Anosov diffeomorphism, which we do not quote.)

Moreover, two special curves go through each point $x \in \mathbb{T}^2$: one, $W^s(x)$, corresponds to the straight line in the eigendirection corresponding to λ_1, and the second line, $W^u(x)$, corresponds to the second eigendirection. Since eigenvalues are irrational, the slope coefficients of the two own directions are irrational. Thus, each of the manifolds $W^s(x)$ and $W^u(x)$ is dense in the torus (creates a skein); they are immersed submanifolds in topological sense.

It turns out that the hyperbolic torus automorphism has the mixing property with respect to the Lebesque measure (which is invariant for T). Moreover, a concrete Markov partitions for this automorphism was constructed by R. Adler and B. Weiss. Finally, we note the following theorem.

Theorem 5.42 (Anosov). *Any Anosov diffeomorphism on a compact manifold is structurally stable.*

Proof. The proof in the general case is given in the book of Z. Nitecki [19]. Here we limit ourselves to the case from the latter example.

Let $U : \mathbb{T}^2 \longmapsto \mathbb{T}^2$ be diffeomorphism close to the diffeomorphism T defined by the matrix A from Eq. (5.19). We can assume that $U(x) = Ax + u(x)$, where the map u is small and doubly periodic. We look for a continuous map $H(x) = x + h(x)$ such that $U \circ H = H \circ T$, i.e., the map h should satisfy the equation

$$h(Ax) - Ah(x) = u(x + h(x)). \tag{5.20}$$

Assuming u small, hence h also small, the map $u(x + h(x)) - u(x)$ is small of second order. Therefore, we firstly solve the equation

$$h(Ax) - Ah(x) = u(x). \tag{5.21}$$

Denoting by L the operator acting on the left hand side on h, we get $h = L^{-1}u$, provided L is invertible.

Let us write $u = u_1 e_1 + u_2 e_2$ and $h = h_1 e_1 + h_2 e_2$, where e_1 and e_2 are unit eigenvectors of the matrix A : $Ae_j = \lambda_j e_j$ with $0 < \lambda_1 < 1 < \lambda_2$. We arrive at the equations

$$(A^* - \lambda_1) h_1 = u_1, \quad (A^* - \lambda_2) h_2 = u_2,$$

where $A^* h(x) = h(Ax)$. We define the inverses as follows:

$$(A^* - \lambda_1)^{-1} = \{A^*(I - \lambda_1 (A^*)^{-1})\}^{-1} = (A^*)^{-1} + \lambda_1 (A^*)^{-2} + \cdots,$$

$$(A^* - \lambda_2)^{-1} = -\{\lambda_2 (I - \lambda_2^{-1} A^*)\}^{-1}$$

$$= -\lambda_2^{-1} - \lambda_2^{-2} A^* - \lambda_2^{-3} (A^*)^2 + \cdots,$$

where the operators $\lambda_1 (A^*)^{-1}$ and $\lambda_2^{-1} A^*$ have norms < 1, as linear operators on the Banach space $C^0 (\mathbb{T}^2, \mathbb{R})$. So, we can solve Eq. (5.21).

We rewrite Eq. (5.20) as $Lh = u \circ (I + h)$ or as

$$h = \mathcal{T}(h),$$

where $\mathcal{T}(h)(x) = L^{-1}(u(x + h(x)))$. The operator $\mathcal{T}$ is contracting, because L^{-1} is bounded and $\sup |u(x + h_1(x)) - u(x + h_2(x))| \leq \sup \|u'\| \cdot \sup |h_1(x) - h_2(x)|$ and $\sup \|u'\|$ is small.

This implies the existence of the map $H = I + h$. We have to show that H is a homeomorphism.

Take the lifts $\widetilde{H} = I + h$, $\widetilde{T} = A$, and $\widetilde{U} = I + u$ as maps on the covering space $\mathbb{R}^2$; they satisfy $\widetilde{H} \circ \widetilde{T} = \widetilde{U} \circ \widetilde{H}$.

Suppose $H(x) = H(y)$. Then $\widetilde{H}(A^n x) = \widetilde{H}(A^n y)$, i.e.,

$$h(A^n x) - h(A^n y) = A^n (y - x).$$

But $A^n (x - y) \to \infty$ as $n \to \infty$ or as $n \to -\infty$, which contradicts the property that h is bounded. So, H is injective.

Next, since h is small the map $\widetilde{H}$ send a sufficiently large ball to a domain containing a large ball. This implies that H is also surjective. $\qquad\square$

Example 5.43 (Geodesic flow on a surface with constant negative curvature). Recall that the **Lobachevskian plane** is the upper half-plane

$$\mathbb{H} = \{\operatorname{Im} z > 0\} \subset \mathbb{C},$$

$z = x + iy$, with the metric

$$\mathrm{d}s^2 = y^{-2} \left(\mathrm{d}x^2 + \mathrm{d}y^2\right). \tag{5.22}$$

Thus, the matrix defining the scalar products is $A = (g_{ij}) = \operatorname{diag}\left(y^{-2}, y^{-2}\right)$. Note also that the angles in this metric are the same as the Euclidean angles. The line $\{y = 0\}$ is called the **absolute**.

According to Section 6.5.2 (in Appendix) the Euler–Lagrange system defining the geodesic flow in this metric is defined by the Lagrangian $L = T = \left(\dot{x}^2 + \dot{y}^2\right)/2y^2$. The corresponding momenta are $p_x = \dot{x}/y^2$ and $p_y = \dot{y}/y^2$; thus

$$\dot{x} = y^2 p_x, \quad \dot{y} = y^2 p_y. \tag{5.23}$$

The additional equations are as follows:

$$\dot{p}_x = 0, \quad \dot{p}_y = -\left(\dot{x}^2 + \dot{y}^2\right)/y^3 = -y\left(p_x^2 + p_y^2\right). \tag{5.24}$$

If we assume

$$T = \frac{y^2}{2}\left(p_x^2 + p_y^2\right) = \frac{1}{2},$$

then systems (5.23)–(5.24) define the geodesic flow $\{g^t\}$ on the three-dimensional manifold

$$M = \{(x, y, p_x, p_y) : y > 0,\ T = 1/2\} = \{(z, v) \in \mathbb{H} \times \mathbb{R}^2 : |v| = 1\}. \tag{5.25}$$

We can find some particular solutions to the above equations. Indeed, assuming $p_x(t) \equiv 0$, we get $x(t) \equiv x_0$. Then the condition

$T = 1/2$ implies $p_y = \pm y$ and we get the equation

$$\dot{y} = y,$$

or $\dot{y} = -y$, with the solution

$$y(y) = y_0 e^t$$

(or $y(t) = y_0 e^{-t}$). This means that the straight half-lines $\{x = x_0, y > 0\}$ are geodesic lines. Note that the absolute $\{y = 0\}$ (including $y = \infty$) is achieved in infinite time.

The metric (5.22) is invariant with respect to the following diffeomorphisms of the half-plane $\mathbb{H}$:

(1) the translations parallel to the absolute, $z \longmapsto z + a$, $a \in \mathbb{R}$;
(2) the dilations $z \longmapsto \lambda z$, $\lambda > 0$;
(3) the conjugation $z \longmapsto -\bar{z}$;
(4) the inversion $z \longmapsto 1/\bar{z}$.

In fact, the maps (3)–(4) imply the invariance with respect to the special Möbius map $z \longmapsto -1/z$. The latter map together with the maps (1) and (2) generate the group $\mathrm{PSL}(2, \mathbb{R}) = \mathrm{SL}(2, \mathbb{R}) / \{\pm I\}$ of Möbius automorphisms of $\mathbb{H}$ of the type

$$z \longmapsto \frac{\alpha z + \beta}{\gamma z + \delta}, \quad \begin{pmatrix} \alpha & \beta \\ \gamma & \delta \end{pmatrix} \in \mathrm{SL}(2, \mathbb{R}). \tag{5.26}$$

The transformations (5.26) send the vertical half-lines either to vertical lines or to half-circles orthogonal to the absolute. Therefore,

The geodesic lines in the Lobachevskian geometry are half-circles orthogonal to the absolute.

There is enough of them to form the complete set of geodesic lines.

There exists another model of the Lobachevskian plane. It is the open unit disk

$$\mathbb{D} = \{|w| < 1\} \subset \mathbb{C}$$

with the metric

$$ds^2 = \left(du^2 + dv^2\right) / (1 - |w|^2)^2, \quad w = u + iv. \tag{5.27}$$

The map $z \longmapsto w = \frac{z-i}{z+i}$ sends $\mathbb{H}$ to $\mathbb{D}$ and metric (5.22) to metric (5.27).

Because of the symmetry of metric (5.27) with respect to the rotations $w \longmapsto we^{i\theta}$, the circles $\{|w| = r\}$, $0 < r < 1$, are the circles in the Lobachevskian geometry with center at $w = 0$, i.e., all its points are equidistant from $w = 0$. By application of the Möbius automorphisms of $\mathbb{D}$ to the latter circles we conclude that all circles in the Lobachevskian geometry of $\mathbb{D}$ are also Euclidean circles. By application of the Möbius relation between $w \in \mathbb{D}$ and $z \in \mathbb{H}$, we conclude the following:

The circles in the metric (5.22) are Euclidean circles, but with the Euclidean centers different from the Lobachevskian centers.

We can consider a sequence of the latter circles which tends to a circle tangent to the absolute. They are called the **horocycles**. It turns out that any horocycle is either a circle tangent to the absolute or a line parallel to the absolute (it is tangent to the absolute at $z = \infty$).

Let us explain more precisely how the horocycles are associated with the given point $(z, v) \in M$. Let ℓ be the geodesic line through z in the direction v. For $t > 0$ take a point $z_t \in \ell$ at a distance t from z and a circle of radius t with center at z_t; this circle passes through z. As $t \to +\infty$ the points z_t tend to z_∞ at the absolute and the circle tends to the horocycle $\Gamma_+ (z, v)$ through z. By taking points $z_{-t} \in \ell$ at the distance t from z and opposite to z_t and circles or radii t with centers at z_{-t} and passing with t to ∞ one gets the other horocycle $\Gamma_- (z, v)$ through z and tangent to the absolute at $z_{-\infty} = \lim z_{-t}$.

If the geodesic line is $\ell = \{x = 0\}$ directed upwards and $z = iy_0$, then the horocycles through z_0 are $\Gamma_+ (iy_0, (0, 1)) = \{y = y_0\}$ and $\Gamma_- (iy_0, (0, 1)) = \{x^2 + (y - y_0/2)^2 = (y_0/2)^2\}$. Note that they are orbits of two 1-parameter subgroups of the group $\mathrm{PSL}\,(2, \mathbb{R})$ of automorphisms of $\mathbb{H}$: one subgroup is defined by the matrices

$$\begin{pmatrix} 1 & s \\ 0 & 1 \end{pmatrix}, \quad s \in \mathbb{R}$$

and the other is defined by

$$\begin{pmatrix} 1 & 0 \\ s & 1 \end{pmatrix}, \quad s \in \mathbb{R}.$$

On the other hand, the geodesic subgroup is defined by

$$\begin{pmatrix} e^{t/2} & 0 \\ 0 & e^{-t/2} \end{pmatrix}, \quad t \in \mathbb{R}.$$

Also for other points $(z, v) \in M$ the horocycles and the geodesic line are orbits of suitable subgroups of $\mathrm{PSL}\,(2, \mathbb{R})$.

Now, we are going to prove that

> The geodesic flow on M satisfies the conditions from Definition 5.33 for hyperbolicity of the invariant set $\Lambda = M$.

Recall that the hyperbolicity condition requires a splitting of the tangent bundle TM into a Whitney sum

$$TM = E^u \oplus E^s \oplus E^0$$

(where E^0 is tangent to the geodesic flow), which is invariant for the derivative maps Dg^t and bounds like in Eq. (5.18) are satisfied.

We define firstly two foliations of M into lines $W^u\,(z, v)$, $W^s\,(z, v)$ such that $E^u_{z,v} = T_{z,v}W^u$ and $E^s_{z,v} = T_{z,v}W^s$. Of course, the subbundle is tangent to the foliation defined by the phase curves if the geodesic flow.

Let $\Gamma_+\,(z, v)$ be the horocycle through z defined by the geodesic $\ell = \ell\,(z, v)$ through z in the direction v. We equip this horocycle with unit vectors orthogonal to it and in the same direction as v. We obtain a 1-parameter family of pairs (ζ, u) such that $\zeta \in \Gamma_+\,(z, v)$ and $u \perp \Gamma_+\,(z, v)$ at ζ, $|u| = 1$. It is one-dimensional submanifold of M, which we claim to be $W^s(z, v)$.

Analogously with the other horocycle $\Gamma_-\,(z, v)$ we associate a submanifold $W^s\,(z, v) \subset M$ defined by unit vectors orthogonal to $\Gamma_-\,(z, v)$ and directed in opposite direction as v.

The fact that the curves $W^u\,(z, v)$, $W^s\,(z, v)$ and $E^0_{z,v}$ are transversal at (z, v) is checked directly; important is the fact that the horocycles have first (not second) order of tangency.

The invariance of the above splitting follows provided the foliations $\{W^u\,(z, v)\}$ and $\{W^s\,(z, v)\}$ are invariant with respect to the geodesic flow. But this means invariance of the horocycles, e.g.,

$$\Gamma_+\,(z(t), v(t)) = \{\xi\,(t) : \zeta\,(0) \in \Gamma_+\,(z, v)\}.$$

This is true when $\Gamma_+\,(z, v)$ and $v = (0, 1)$ is a horizontal line. General case can be reduced to this one; we omit details.

Finally, we will show that segments of the horocycle are contracted exponentially under the action of the geodesic flow; this suffice to show estimates (5.18). Again, assume $\Gamma_+ = \{y = y_0\}$. Here the geodesic lines are $\{x = \text{const}\}$ and the geodesic flow sends the line $\{y = y_0\}$ to the line $\{y = y_t = y_0 e^t\}$. Thus, each segment $I \subset \{y = y_0\}$ is sent to a parallel segment I_t in $\{y = y_0 e^t\}$. But the Lobachevskian length $|I_t| = (y_0/y_t)|I| = e^{-t}|I|$.

Therefore,

The geodesic flow on M is an Anosov flow.

This result can be applied to geodesic flows on compact Riemann surfaces of constant negative curvature. Such a surface S, after some rescaling of its area, is obtained from $\mathbb{H}$ by applying an action of a discrete subgroup of its automorphism, $S = \mathbb{H}/G$. Recall that S has genus $g \geq 2$ and topologically it is a 2-sphere with g handles attached. After cutting it along $2g$ curves one obtains a curvilinear polygon P with $4g$ sides. This polygon is embedded into the hyperbolic plane $\mathbb{H}$ in suitable way (we do not present details) from which it borrows the hyperbolic metric. The phase space of the geodesic flow on is $M_S = \{(z,v) : z \in S, v \in T_z S\}$ and it is a quotient of M by an action of a suitable discrete group. It can be treated as

$$M_P = \{(z,v) : z \in P, |v| = 1\}$$

with some sides identified.

The foliations of M into phase lines of the geodesic flow and the stable and unstable manifolds gives foliations of M_P which behave well after identifications. So, we obtain suitable foliations on M_S which imply that the geodesic flow is hyperbolic. Therefore,

The geodesic flow on a compact Riemann surface with constant negative curvature is an Anosov flow.

Remark 5.44. The concept of Anosov diffeomorphism (and of Anosov flow) looks attractive, because of its global character. But the examples are rather scarce.

For example, A. Manning proved that any Anosov diffeomorphism on a torus is conjugated with a linear hyperbolic automorphism of $\mathbb{T}^n = \mathbb{R}^n/\mathbb{Z}^n$ defined by a hyperbolic matrix A with integer entries and with $\det A = 1$.

5.6.3. *Attractors*

Very important examples of dynamic systems are the so-called **hyperbolic attractors**. These are smooth transformations (not necessarily invertible) $T : M \longmapsto M$ for which there exists a closed invariant subset $\Lambda \subset M$ with a neighborhood $U \supset \Lambda$ such that $\Lambda = \bigcap_{n \geq 0} T^n(U)$. Locally, Λ has the form $N \times C$, where N is a regular manifold (with $0 < \dim N < \dim M$) and C is the Cantor set.

Moreover, Λ has a hyperbolic structure in the sense that $T_*(x)$ uniformly expands in the direction of N and uniformly contracts in the transversal to N direction.

Example 5.45 (Selenoid). Let $M = D^2 \times \mathbb{S}^1 = \{(z,y)\}$ be a full torus, where $D^2 = \{z : |z| \leq 1\} \subset \mathbb{C}$ is the disk and $\mathbb{S}^1 = \{y \mod \mathbb{Z}\}$. The transformation is given by

$$T : (z,y) \longmapsto \left(\frac{1}{4}z + \frac{1}{2}\mathrm{e}^{2\pi i y}, 2y \mod \mathbb{Z} \right).$$

The image $T(M) \subset M$ will be four times thinner and twice longer and inserted in M so that it wraps twice around the "equator" M while slightly twisting (see Fig. 5.11).

Of course, $\Lambda = \bigcap_{n \geq 0} T^n(M)$ is an invariant set and satisfies the requirements that were imposed above on hyperbolic attractors.

Finally, we would like to point out that in the Dynamic Systems Theory there exist the so-called *strange attractors* whose dynamics is difficult to describe. Strange attractors satisfy the property $\Lambda = \bigcap_{n \geq 0} T^n(U)$, but do not want to be uniformly hyperbolic.

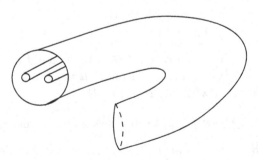

Fig. 5.11. Selenoid.

The best-known are the **Hénon map**, given by

$$(x, y) \longmapsto \left(y + 1 - ax^2, bx\right)$$

(where, e.g., $a = 1.4$ and $b = 0.3$), and the **Lorenz attractor**, given the by vector field

$$\dot{x} = -\sigma x + \sigma y, \quad \dot{y} = -xz + rx - y, \quad \dot{z} = xy - bz$$

(where, e.g., $\sigma = 10$, $r = 28$ and $b = 8/3$).

Finally, in the book of Z. Nitecki [19] the reader may find more details on the concepts of this section.

5.7. Exercises

Exercise 5.46. Prove that any orientation-reversing diffeomorphism of the circle has exactly two distinct fixed points and any other periodic point has period 2.

Exercise 5.47. Prove that, if f and h are orientation-preserving diffeomorphisms, then $\tau(f) = \tau\left(h \circ f \circ h^{-1}\right)$.

Exercise 5.48. Let $f : \mathbb{S}^1 \longmapsto \mathbb{S}^1$ be given orientation-preserving diffeomorphism and $R_\alpha(x) = x + \alpha \mod 1$ be the rotation by the angle α. Prove that the function $\alpha \longmapsto \tau(R_\alpha \circ f)$ is continuous. *Hint*: Use a suitable lift.

Exercise 5.49. Prove that diffeomorphisms with rational rotation number and only hyperbolic periodic points are structurally stable. Prove also that the subset in $C^1\left(\mathbb{S}^1, \mathbb{S}^1\right)$ of such diffeomorphisms is open and dense (the Kupka–Smale theorem). *Hint*: Use the previous exercise.

Exercise 5.50. Prove that, if a diffeomorphism f of the circle preserves an absolutely continuous measure $\mu = \rho(x)\, \mathrm{d}x$, then f is topologically conjugated with the rotation $R_{\tau(f)}$.

Exercise 5.51. Draw $f^2(A)$ in the case of the Smale transformation.

Exercise 5.52. Show that $f^{-n}(A) \cap A$, $n > 0$, consists of 2^n horizontal strips.

Exercise 5.53. Prove that Λ from Eq. (5.12) is invariant.

Exercise 5.54. Show that Λ (in Eq. (5.12)) is homeomorphic with $C \times C$, where C is the (properly defined) Cantor set.

Exercise 5.55. Check that Φ conjugates f with σ.

Exercise 5.56. Prove that, if a sequence $\{a_n\}_{n=1,2,\ldots}$ of positive numbers satisfies the inequality $a_{m+n} \leq a_m + a_n$, then the sequence $\{a_n/n\}$ is convergent.

Exercise 5.57. Prove that the transitive translation on the torus is not mixing and has zero entropy.

Exercise 5.58. Prove that the set of periodic points of the map from Example 5.41 coincides with the set of points with both coordinates rational.

Hint: The set $\{(p/N, q/N) \mod \mathbb{Z}^2 : p, q \in \mathbb{N}\}$ for given $N \in \mathbb{N}$ is finite and invariant for f. Moreover, the equations for the periodic point of the period n take the form $(A^n - I)x = m$, where $m = (m_1, m_2) \in \mathbb{Z}^2$.

Chapter 6

Appendix: Basic Concepts and Theorems of the Theory of ODEs

6.1. Definitions

Under the **ordinary differential equation** (ODE) we understand an equation of the form

$$\frac{\mathrm{d}x}{\mathrm{d}t} = \dot{x} = v(t, x), \tag{6.1}$$

where $t \in I \subset \mathbb{R}$ is real time (I is an open interval), x belongs to the phase space (manifold) M and v is a time-dependent vector field on M, $v : I \times M \longmapsto TM$ such that $v(t, x) \in T_x M$. Often, $M = U$ is an open subset of the Euclidean space $\mathbb{R}^n$; then $v : I \times U \longmapsto \mathbb{R}^n$ and we are talking about a system of ordinary differential equations. If v does not depend on time, $v = v(x)$, then Eq. (6.1) is an **autonomous equation** (and v is an **autonomous vector field**), otherwise we are dealing with a non-autonomous equation. The space $I \times M$ is called the **extended phase space**.

We call the **solution** of Eq. (6.1) a differentiable map $\varphi : J \longmapsto M$, $J \subset I$, which satisfies the equation

$$\frac{\mathrm{d}\varphi}{\mathrm{d}t}(t) \equiv v(t; \varphi(t)).$$

We call the **initial value problem** the following two conditions:

$$\dot{x} = v(t, x), \ x(t_0) = x_0, \tag{6.2}$$

233

the second of which is called the **initial condition**. The **solution to the initial value problem** (6.2) is the solution

$$\varphi(t) = \varphi(t; x_0, t_0)$$

of Eq. (6.1), which has the property $\varphi(t_0) = x_0$.

If $\varphi(t)$ is a solution of system (6.1), then the curve $\{(t, \varphi(t)) : t \in J\} \subset I \times M$ (i.e., the graph of the solution) is called the **integral curve**; if, additionally, the system (6.1) is autonomous, then the curve $\{\varphi(t) : t \in J\}$ (i.e., the image of the solution) is called the **phase curve**.

Remark 6.1. By introducing a new time τ, we can rewrite the non-autonomous equation (6.1) in the form of the following autonomous system:

$$\frac{dt}{d\tau} = 1, \quad \frac{dx}{d\tau} = v(t, x) \tag{6.3}$$

in the extended phase space. Then the integral curves for Eq. (6.1) will turn out to be phase curves for system (6.3).

The *differential equation of the order n*, i.e.,

$$\frac{d^n x}{dx^n} = x^{(n)} = f(t, x, x^{(1)}, \dots, x^{(n-1)}), \ t \in I, \ x \in \mathbb{R} \tag{6.4}$$

is replaced by the system of first-order equations

$$\dot{y}_1 = y_2, \ \dot{y}_2 = y_3, \dots, \dot{y}_{n-1} = y_n, \ \dot{y}_n = f(t, y_1, \dots, y_n) \tag{6.5}$$

by the substitution $x = y_1$, $x^{(1)} = y_2, \dots, x^{(n-1)} = y_n$. The natural initial condition for Eq. (6.4) is

$$x(t_0) = x_0, \ x^{(1)}(t_0) = x_1, \dots, x^{(n-1)}(t_0) = x_{n-1}. \tag{6.6}$$

Note that, by using the trick from Remark 6.1, we can replace (in general) the non-autonomous system (6.5) with an appropriate autonomous system in $\mathbb{R}^{n+1}$.

Remark 6.2. In books on differential equations, the *implicit equations with respect to the derivative*, of the type

$$F(t, x, \dot{x}) = 0, \ t \in \mathbb{R}, \ x \in \mathbb{R} \tag{6.7}$$

are also considered. It turns out that, if the equation $F(t, x, p) = 0$ can be solved around a certain point (t_0, x_0, p_0) in the form of

$x = g(t, p)$, then Eq. (6.7) can be rewritten as an autonomous system

$$\frac{dt}{d\tau} = g'_p(t, p), \quad \frac{dp}{d\tau} = p - g'_t(t, p),$$

where τ is the new "time". Indeed, we have $\frac{dx}{d\tau} = \frac{dx}{dt}\frac{dt}{d\tau} = p\frac{dt}{d\tau}$. Thus, by differentiating the identity $x(\tau) = g(t(\tau), p(\tau))$, we get the condition $p\frac{dt}{d\tau} \equiv g'_t\frac{dt}{d\tau} + g'_p\frac{dp}{d\tau}$. It is satisfied for the above vector field.

A similar system can be written when the equation $F = 0$ is solved with respect to t, and also when $x \in \mathbb{R}^n$ and $F \in \mathbb{R}^n$. In this book, equations of type (6.7) are not studied, but we cited them to demonstrate some universality of autonomous differential equations.

The concept of a phase flow is connected with the autonomous equation

$$\dot{x} = v(x). \tag{6.8}$$

Note that the solutions $\varphi(t; x_0, 0)$ of the Eq. (6.8) with the initial condition $x(0) = x_0$ define the family of maps

$$g^t : D_t \longmapsto M, \quad x_0 \longmapsto \varphi(t; x_0, 0),$$

where D_t is the domain of definition of the mapping g^t. This family should fulfill two-natural properties:

$$g^0 = id, \tag{6.9}$$

$$g^t \circ g^s = g^{t+s}. \tag{6.10}$$

Property (6.9) is the definition of the initial condition. Property (6.10), which should be satisfied for $x_0 \in D_s \cap (g^s)^{-1}(D_t)$, means that if we start at time 0 from point x_0 and we arrive (along the solution) at the point $y_0 = g^s(x_0)$ at time s and then reset the stopwatch and go from y_0 after time t, then we come to that same point as if we were going after the time $t + s$ from x_0 without resetting the stopwatch. Of course, here it is important that $v(s, y_0) = v(0, y_0) = v(y_0)$ (autonomy). A family $\{g^t\}_{t\in I}$, $g^t : D_t \longmapsto M$ which satisfies the properties (6.9)–(6.10) is called the **local phase flow**. A family

$$g^t : M \longmapsto M, \quad t \in \mathbb{R},$$

of (global) diffeomorphisms of the phase space M, satisfying properties (6.9)–(6.10) is called the **phase flow** on M. In other words, the mapping $t \longmapsto g^t$ is a homomorphism from the group $\mathbb{R}$ to the group $\mathrm{Diff}(M)$ of the diffeomorphisms of the manifold M.

Example 6.3. The equation

$$\dot{x} = x^2 + 1$$

defines a global vector field on the projective space $\mathbb{RP}^1 = \mathbb{R} \cup \infty$ (where the coordinate $y = 1/x$, in a neighborhood of $x = \infty$, satisfies the equation $\dot{y} = -1 - y^2$). Here the local phase flow turns out to be a phase flow on $\mathbb{RP}^1$ composed of Möbius transformations

$$g^t(x_0) = \frac{x_0 \cos t + \sin t}{\cos t - x_0 \sin t}.$$

Remark 6.4. In the case of a non-autonomous vector field, we deal with a 2-parameter family of transformations

$$g_s^t : M \longmapsto M$$

(more precisely, with its local version) defined in such a way that $g_s^t(x_0) = \varphi(t; x_0; s)$, i.e., the value at the time t of the solution starting from x_0 at the moment s. There are obvious identities

$$g_t^t = id, \ g_s^t \circ g_u^s = g_u^t.$$

6.2. Theorems

In the following, the reader will find a series of theorems which are fundamental in the theory of ODEs and which are given without complete proofs. For more details we refer the reader to [1,5,7,8,20–22].

Theorem 6.5 (On existence and uniqueness of local solutions).

Let us suppose that the field $v(t, x)$ is of class C^1 on an open set $I \times U \subset \mathbb{R} \times \mathbb{R}^n$. Let $(t_0, x_*) \in I \times U$.

Then there exists an interval $I_0 \subset I$, containing the starting moment t_0, and a neighborhood $U_0 \subset U$ of point x_* such that for any $x_0 \in U_0$ the initial value problem $\dot{x} = v(t; x)$, $x(t_0) = x_0$, has exactly one solution $\varphi(t; x_0) = \varphi(t; x_0, t_0)$.

Moreover, the mapping

$$(t, x_0) \longmapsto \varphi(t; x_0) \tag{6.11}$$

is continuous, and, if the field $v(t, x)$ is analytic, then this mapping is also analytic.

We recall that the basic idea of proof of this theorem consists in replacing the initial value problem (6.2) with the integral equation

$$\varphi(t; x_0) = x_0 + \int_{t_0}^t v(t, \varphi(s; x_0)) \mathrm{d}s. \tag{6.12}$$

This equation is treated as a fixed point equation $\varphi = \mathcal{T}(\varphi)$ for the operator defined on the right-hand side of Eq. (6.12) acting on an appropriate Banach space of mappings $\varphi(t, x_0)$. In general, this is the Banach space of continuous (vector valued) functions on $I_0 \times U_0$ with the supremum norm, while the contraction property of the operator $\mathcal{T}$ results from the Lipschitz condition with respect to x for the field $v(t, x)$. In the analytic case, the chosen Banach space is the space of holomorphic functions in a certain domain in $\mathbb{C} \times \mathbb{C}^n$ with the supremum norm (Exercise 6.38).

Example 6.6. The equation

$$\dot{x} = \sqrt[3]{x}$$

has two solutions with the same initial condition $x(0) = 0$: $\varphi_1(t) = 0$ for $t < 0$ and $\varphi_1(t) = (2t/3)^{3/2}$ for $t \geq 0$ and $\varphi_2(t) \equiv 0$. This standard example shows how important the Lipschitz condition is; here it fails at $x = 0$.

Theorem 6.7 (On dependence on the initial condition).

If, in Theorem 6.5, we assume that v is of class C^2, then the mapping (6.11) will be of class C^1. More generally, if v is of class C^r, $1 \leq r \leq \infty$, then φ is of class C^{r-1}.

Theorem 6.8 (On the dependence on parameters).[a] *If the field v depends on an additional parameter $\lambda \in V \subset \mathbb{R}^k$ and $v(t, x; \lambda)$ is of class C^r, $r \geq 2$, then the solution $\varphi(t; x_0; \lambda)$ is of class C^{r-1}.*

The proofs of the last two theorems use the notion of equation in variations.

The **equation in variations with respect to initial condition** is the equation

$$\dot{y} = A(t)y, \quad A(t) = \frac{\partial v}{\partial x}(t, \varphi_0(t)). \tag{6.13}$$

Here $\varphi_0(t)$ is a given solution, with the initial condition $\varphi_0(t_0) = x_0$, and Eq. (6.13) is obtained by substituting the perturbation $x = \varphi_0(t) + \varepsilon y(t) + O(\varepsilon^2)$ (with small ε) to the initial value problem (6.2) with the perturbed initial condition $x(t_0) = x_0 + \varepsilon y_0$ and comparison of terms of order ε. The partial derivative $\partial \varphi / \partial (x_0)_j$ of the solution $\varphi(t; x_0, t_0)$ with respect to the initial condition is obtained as a solution of system (6.13) with the initial condition $y_0 = e_j$ (where (e_j) is the standard base in $\mathbb{R}^n$).

The **equation in variations with respect to parameter** is the equation

$$\dot{y} = A(t)y + b(t), \quad b(t) = \frac{\partial v}{\partial \lambda}(t, \varphi_0(t); \lambda_0). \tag{6.14}$$

Here $\varphi_0(t)$ is a distinguished solution to the initial value problem $\dot{x} = v(t, x, \lambda_0)$, $x(t_0) = x_0$, i.e., for a fixed parameter $\lambda = \lambda_0$, and the matrix $A(t)$ is same as in Eq. (6.13). This equation is obtained by substituting $x = \varphi_0(t) + \varepsilon y(t) + O(\varepsilon^2)$ to the initial value problem $\dot{x} = v(t, x; \lambda_0 + \epsilon \nu_0)$, $x(t_0) = x_0$, and comparing the linear terms with respect to the small ε.

In the proofs of Theorems 6.7 and 6.8, the problem boils down to considering the system $\dot{x} = v(t, x)$, $\dot{y} = \frac{\partial v}{\partial x}(t, x)y$, or the system $\dot{x} = v(t, x; \lambda)$, $\dot{y} = \frac{\partial v}{\partial x}(t, x; \lambda)y + \frac{\partial v}{\partial \lambda}(t, x; \lambda)$, and applying Theorem 6.5 (Exercises 6.39 and 6.40).

From the above theorems important conclusions about the qualitative behavior of solutions of Eq. (6.1) follow.

---

[a]Some sources (e.g. [22]) show the C^r dependence of solutions on the parameters. For our purposes, the class C^{r-1} is sufficient, especially when considering the simplicity of the following sketch of the proof of this theorem.

Theorem 6.9 (Flow box for non-autonomous systems). *If* $v(t,x)$ *is of class* C^r, $r \geq 2$, *and* $(t_0, x_*) \in I \times U \subset \mathbb{R} \times \mathbb{R}^n$, *then there exists a local diffeomorphism*

$$f : (t,x) \longmapsto (t,y)$$

from a neighborhood of the point (t_0, x_*), *which transforms system* (6.1) *to the system*

$$\dot{y} = 0.$$

In the proof, the diffeomorphism f is defined such that, if the point x equals $x = \varphi(t; x_0, t_0)$, i.e., it is the value of the solution with initial condition $x(t_0) = x_0$ after time t, then we put $y = x_0$ (Exercise 6.41).

Theorem 6.10 (Flow box for autonomous systems). *If an autonomous vector field* $v(x)$ *is of class* C^r, $r \geq 2$, *on* U *and a point* $x_* \in U$ *is such that*

$$v(x_*) \neq 0, \tag{6.15}$$

there exists a local diffeomorphism $f : x \longmapsto y$ *from a neighborhood of point* x_*, *which transforms the system* $\dot{x} = v(x)$ *to the system*

$$\dot{y}_1 = 1, \quad \dot{y}_2 = 0, \ldots, \quad \dot{y}_n = 0.$$

As can be guessed, the variable y_1 is the time t of solutions $\varphi(t; x_0)$, which starts at $t = 0$ from a hyperplane H perpendicular to the vector $v(x_*)$. The remaining variables y_j are derived from some coordinate system on the hyperplane H and are constant along the solutions (Exercise 6.42).

Remark 6.11. The above theorems can be called the first theorems of the Qualitative Theory of ODEs. They say that, locally, all vector fields (satisfying condition (6.15) in the autonomous case) are the same from a mathematical point of view. Condition (6.15) implies a certain simplicity of the autonomous vector field. In the first chapter of this book, we study the situation when this condition is violated.

Theorem 6.12 (About local phase flow). *For an autonomous vector field* $v(x)$ *of class* C^r, $r \geq 2$, *there is a local phase flow* $\{g^t\}$, $x_0 \longmapsto g^t(x_0)$ *(satisfying conditions* (6.9)–(6.10)*) given by solutions* $\varphi(t; x_0)$ *of initial value problem* $\dot{x} = v(x)$, $x(0) = x_0$.

Of course, this theorem is an immediate consequence of the theorem about existence and uniqueness of local solutions for problem (6.1) with the autonomous field $v(x)$.

Theorem 6.13 (About prolongation of solutions). *Let the field $v(t, x)$ be of class C^r, $r \geq 1$, in the open set $I \times U$ and let $F \subset U$ be a compact subset. Then any local solution $\varphi(t; x_0; t_0)$ starting from $x_0 \in F$ either is extended for all times $t_0 \leq t < \infty$ while remaining in F, or goes out of F after a finite time $T(x_0) \geq t_0$.*

The same alternative takes place for solutions $\varphi(t; x_0; t_0)$ with $t \leq t_0$.

In a sense, this theorem is obvious. The following example shows that the compactness assumption of F is important.

Example 6.14. The equation

$$\dot{x} = x^2, \quad x \in \mathbb{R},$$

has solutions $\varphi(t) = x_0/(1 - tx_0)$, which escape to infinity after the finite time $T = 1/x_0$.

6.3. Methods of solving

Following is a list of classes of ordinary differential equations, which can be integrated, and methods for their integration are given. All the equations considered here have the form

$$\frac{\mathrm{d}y}{\mathrm{d}x} = \frac{Q(x, y)}{P(x, y)}, \tag{6.16}$$

or the equivalent form of the *Pfaffian equation*

$$Q(x, y)\mathrm{d}x - P(x, y)\mathrm{d}y = 0.$$

Example 6.15 (Equations with separated variables). These are equations of the form

$$\frac{\mathrm{d}y}{\mathrm{d}x} = \frac{Q(x)}{P(y)}.$$

Of course, the solutions are given in the implicit form

$$\int_{x_0}^{x} Q(z)\mathrm{d}z = \int_{y_0}^{y} P(y)\mathrm{d}y.$$

Example 6.16 (Homogeneous equations). These are of the form

$$\frac{dy}{dx} = f(y/x).$$

Here the substitution $u = \frac{y}{x}$ leads to the equation

$$x\frac{du}{dx} = f(u) - u,$$

with separated variables. This class may include equations of the form

$$\frac{dy}{dx} = f\left(\frac{ax + by + \alpha}{cx + dy + \beta}\right), \quad \begin{vmatrix} a & b \\ c & d \end{vmatrix} \neq 0.$$

By moving the origin of the coordinate system to the point of intersection of the lines $ax + by + \alpha = 0$ and $cx + dy + \beta = 0$, it becomes evidently homogeneous. When $ad - bc = 0$, the equation is easily reduced to an equation with separated variables.

Example 6.17 (Quasi-homogeneous equations). Such equations are characterized by invariance with respect to a symmetry of the type

$$x \longmapsto \lambda x, \ y \longmapsto \lambda^\gamma y, \quad \lambda \in \mathbb{R} \setminus 0,$$

which generalizes analogous symmetry with $\gamma = 1$ for homogeneous equations. Here the substitution $u = y/x^\gamma$ leads to an equation with separated variables.

Example 6.18 (Linear equations). The linear equations

$$\frac{dy}{dx} = a(x)y + b(x) \qquad (6.17)$$

are divided into *homogeneous* ones, when $b(x) \equiv 0$, and *non-homogeneous*. The general solution of the homogeneous equation $\frac{dy}{dx} = a(x)y$, associated with Eq. (6.17), has the form

$$\varphi_{\text{gen}} = C \cdot \exp A(x),$$

where $A(x)$ is a primitive function for the function $a(x)$. A general solution of the non-homogeneous equation is the sum of a general solution of the homogeneous equation φ_{gen} and a particular solution φ_{part} of the non-homogeneous equation. We are looking for the latter solution using the *variation of constants method*, i.e., in the form

$$\varphi_{\text{part}} = C(x) \cdot \exp A(x).$$

After substituting to Eq. (6.17), we obtain the equation $C'(x) = e^{-A(x)}b(x)$.

The general solution is

$$y = e^{A(x)}C + \int^x e^{A(x)-A(z)}b(z)\mathrm{d}z. \tag{6.18}$$

Example 6.19 (Bernoulli equation).

$$\frac{\mathrm{d}y}{\mathrm{d}x} = a(x)y + b(x)y^n$$

is reduced to a linear equation by substituting

$$z = y^{1-n}.$$

Example 6.20 (Equations with an integrating factor). An equation with an integrating factor is as follows:

$$\frac{\mathrm{d}y}{\mathrm{d}x} = \frac{Q(x,y)}{P(x,y)} = \frac{-MH'_x}{MH'_y},$$

or

$$M(H'_x\mathrm{d}x + H'_y\mathrm{d}y) = M\mathrm{d}H = 0.$$

Here $M = M(x,y)$ is the **integrating factor** and $H = H(x,y)$ is the **first integral** of the equation; thus, the function H is constant on the integral curves of the equation, $H(x, \varphi(x)) \equiv \text{const.}$ Of course, here the solutions $y = \varphi(x)$ are given implicitly in the form

$$H(x,y) = h.$$

The natural question is how to guess from the form of the functions P and Q if there is an integrating factor and the first integral. It is convenient to work with corresponding autonomous vector field

$$\dot{x} = P(x,y), \ \dot{y} = Q(x,y), \tag{6.19}$$

associated with Eq. (6.16).

Note that the case of $M(x, y) \equiv 1$, with the first integral $H(x, y)$, corresponds to the situation when system (6.19) is *Hamiltonian* with H as the *Hamilton function* (*Hamiltonian*),

$$\dot{x} = H'_y, \ \dot{y} = -H'_x.$$

Of course, then we have

$$\operatorname{div} V = P'_x + Q'_y \equiv 0, \tag{6.20}$$

i.e., the divergence of the vector field $V = Q\partial/\partial x + P\partial/\partial y$ vanishes, or, equivalently,

$$\mathrm{d}\,(Q\mathrm{d}x - P\mathrm{d}y) = 0.$$

It is the necessary condition for the system (6.19) for being Hamiltonian.

When $\operatorname{div} V \equiv 0$, one can define the function H as follows:

$$H(x, y) = \int_{\Gamma(x.y)} Q\mathrm{d}x - P\mathrm{d}y,$$

where $\Gamma(x, y)$ is the path from a fixed point (x_0, y_0) to (x, y). If the domain $U \subset \mathbb{R}^2$, in which system (6.19) is defined, is simply connected (each loop can be contracted to a point), then the definition of $H(x, y)$ does not depend on the choice of path $\Gamma = \Gamma(x, y)$: the difference between this value and the value defined for another path Γ' is the integral along the closed loop $\Gamma - \Gamma'$ (which bounds a domain Ω) of the 1-form $\omega = Q\mathrm{d}x - P\mathrm{d}y$, which is closed, so the Stokes formula gives $\oint_{\Gamma - \Gamma'} \omega = \iint_\Omega \mathrm{d}\omega = 0$ (see Fig. 6.1).

The example

$$\mathrm{d}\left(\operatorname{arctg}\frac{y}{x}\right) = \frac{-y}{x^2 + y^2}\mathrm{d}x + \frac{x}{x^2 + y^2}\mathrm{d}y = 0$$

in $\mathbb{R}^2 \backslash 0$, which fulfills condition (6.20), and has a local (but not global) first integral $H = \arg{(x + iy)}$ shows that the assumption of simply connectivity is significant.

The case when there is a non-trivial integrating factor M is much more difficult. Let us quote here a result of M. Singer, which concerns the case when P and Q are polynomials.

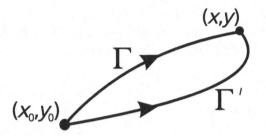

Fig. 6.1.   Definition of the first integral.

Theorem 6.21 (Singer). *If system* (6.19) *with polynomials P and Q has a first integral, which can be expressed in quadratures, then the integral factor M can be chosen in the so-called Darboux form*

$$M = e^{g(x,y)} f_1^{a_1}(x,y) \ldots f_r^{a_r}(x,y),$$

where $g(x,y)$ is a rational function, $f_j(x,y)$ are polynomials and $a_j \in \mathbb{C}$.[b]

6.4.   Linear systems and equations

Linear systems of ordinary differential equations are generalizations of Eq. (6.17) and have the form

$$\dot{x} = A(t)x + b(t), \quad t \in I \subset \mathbb{R}, \quad x \in \mathbb{R}^n. \tag{6.21}$$

In parallel, we consider **linear differential equations of the order** n of the form

$$x^{(n)} + a_{n-1}(t)x^{(n-1)} + \cdots + a_0(t)x = b(t), \quad t \in I \subset \mathbb{R}, \quad x \in \mathbb{R}. \tag{6.22}$$

It is known that the solutions $x = \varphi(t; x_0, t_0)$ of such systems and equations extend to the whole interval I (Exercise 6.46).

In the homogeneous case, i.e., when $b(t) \equiv 0$, the set of solutions forms an n-dimensional vector space. Each base of this space creates the so-called **fundamental system** $(\varphi_j)_{j=1,\ldots,n}$. Such a fundamental

---

[b]We refer the reader to the monograph *The Monodromy Group* by the first author, where one can find a definition of functions represented in quadratures and a proof of the Singer theorem.

system defines the **fundamental matrix** $\mathcal{F}(t) = (\varphi_1, \ldots, \varphi_n)$ in the case of system (6.21) and

$$\mathcal{F}(t) = \begin{pmatrix} \varphi_1 & \cdots & \varphi_n \\ \varphi_1^{(1)} & \cdots & \varphi_n^{(1)} \\ \vdots & \ddots & \vdots \\ \varphi_1^{(n-1)} & \cdots & \varphi_n^{(n-1)} \end{pmatrix}$$

in the case of Eq. (6.22). The determinant of the fundamental matrix is called the **Wronskian**

$$W(t) = \det \mathcal{F}(t) \tag{6.23}$$

(from the name of the Polish mathematician J. Hoene-Wroński).

A general solution of the homogeneous system equation 6.21 (with $b \equiv 0$) has the form

$$\varphi(t) = \mathcal{F}(t) \cdot C, \tag{6.24}$$

where C is a constant vector (determined from the initial conditions); in particular, when the fundamental system is chosen such that $\mathcal{F}(t_0) = I$, then the solution $\varphi(t) = \mathcal{F}(t)x_0$ satisfies the initial condition $\varphi(t_0) = x_0$.

In the case of homogeneous equation (6.22) (with $b \equiv 0$), the general solution has the form

$$\varphi(t) = (\mathcal{F}(t) \cdot C)_1 = C_1\varphi_1(t) + \cdots + C_n\varphi_n(t),$$

i.e., the first component of the vector standing on the right-hand side of Eq. (6.24).

It is not difficult to guess that the general solution of a non-homogeneous system or equation (i.e., with $b(t) \not\equiv 0$) is the sum of the general solution of the homogeneous equation φ_{gen} and a particular solution of the non-homogeneous system or equation φ_{part}. To solve a non-homogeneous system or equation, knowing the fundamental matrix, we use the *method of variation of constants*, i.e., we substitute $x = \mathcal{F}(t) \cdot C(t)$. By solving the appropriate equation for $C(t)$, we find the general solution of system (6.21) in the form

$$x = \mathcal{F}(t)C + \int_{t_0}^{t} \mathcal{F}(t)\mathcal{F}^{-1}(s)b(s)ds.$$

Of course, the main problem is to find the fundamental matrix $\mathcal{F}(t)$.

In the case when the matrix $A(t) = A$ in system (6.21), or the coefficients $a_j(t) = a_j$ in Eq. (6.22), do not depend on time, we are talking about a *system with constant coefficients* or an *equation with constant coefficients*. In this case, the fundamental matrix has the form

$$\mathcal{F}(t) = \exp At = I + At + \frac{t^2}{2!}A^2 + \cdots,$$

where

$$A = \begin{pmatrix} 0 & 1 & 0 & \cdots & 0 \\ 0 & 0 & 1 & \cdots & 0 \\ \vdots & \vdots & \vdots & \ddots & \vdots \\ -a_0 & -a_1 & -a_2 & \cdots & -a_{n-1} \end{pmatrix}$$

in the case of Eq. (6.22).

For Eq. (6.22) with constant coefficients, the general solution of the homogeneous equation can be obtained directly from the *characteristic equation*

$$P(\lambda) = \lambda^n + a_{n-1}\lambda^{n-1} + \cdots + a_0 = 0. \tag{6.25}$$

It has the form

$$\begin{aligned} \varphi_{\text{gen}}(t) &= (C_{1,0} + C_{1,1}t + \cdots + C_{1,k_1-1}t^{k_r-1})e^{\lambda_1 t} + \cdots \\ &\quad + (C_{r,0} + \cdots + C_{r,k_r-1}t^{k_r-1})e^{\lambda_r t}, \end{aligned} \tag{6.26}$$

where λ_j are the roots of the characteristic equation of multiplicity k_j; in the case of complex pairs of roots $\lambda_j = \bar{\lambda}_{j+1} = \alpha_j + i\beta_j$, corresponding coefficients in the sum (6.26) are conjugated, $C_{j+1,l} = \overline{C_{j,l}}$, and these two summands give the expression

$$D_{j,l}t^l e^{\alpha_j t}\cos(\beta_j t) + E_{j,l}t^l e^{\alpha_j t}\sin(\beta_j t)$$

(with constants $D_{j,l}$ and $E_{j,l}$).

There is also a special recipe for the solution of the non-homogeneous equation (6.22) with constant coefficients, in case when the function $b(t)$ (on the right side of the equation) is a so-called *quasi-polynomial* of the form

$$b(t) = e^{\mu t}p(t). \tag{6.27}$$

Here μ is called the *exponent* of the quasi-polynomial and $p(t)$ is a polynomial of degree m, called the *degree* of quasi-polynomial. Also the functions of the form $e^{\nu t}\cos(\xi t)p(t)$ and $e^{\nu t}\sin(\xi t)p(t)$ are, respectively, the real and imaginary part of a quasi-polynomial with the complex exponent $\mu = \nu + i\xi$.

Theorem 6.22. *The general solution of the homogeneous equation has the form* (6.26).

If the right-hand side of the non-homogeneous equation (6.22) *has the form* (6.27) *and the exponent μ of the quasi-polynomial is a root of the characteristic equation* (6.25) *of multiplicity k, then a particular solution of the equation can be chosen as a quasi-polynomial*

$$\varphi_{\text{part}} = t^k e^{\mu t} q(t),$$

where $q(t)$ is polynomial of degree $m = \deg p(t)$.

The following theorem, due to J. Liouville, is a generalization of the elementary algebraic identity

$$\det \exp A = \exp \operatorname{tr} A$$

and has a huge application in the Qualitative Theory.

Theorem 6.23 (Liouville). *The Wronskian $W(t)$ associated with a fundamental matrix $\mathcal{F}(t)$ of the system* (6.21) *(formula* (6.23)*) satisfies the equation*

$$\dot{W} = \operatorname{tr} A(t) \cdot W.$$

The proof is reduced to finding the limit

$$\lim_{s \to 0} \frac{\det(I + sA(t)) - 1}{s},$$

since $\mathcal{F}(t + s) = (I + sA(t))\mathcal{F}(t) + O(s^2)$. It is easy to check, using the standard definition of determinant $\det(I + sA) = \sum (-1)^\pi \prod (I + sA)_{j,\pi(j)}$, that the summands arising from non-trivial permutations π give a contribution of order s^2. The term $\prod (I + sA)_{j,j} = \prod(1 + sa_{jj})$ is $1 + s\sum a_{jj} + O(s^2)$.

In the case when the fundamental matrix $\mathcal{F}(t)$ satisfies the property $\mathcal{F}(t_0) = I$, the Wronskian determinant has the natural interpretation of the (n-dimensional) volume of the parallelepiped spanned by the vectors $f_i(t) = g_{t_0}^t e_i$, $i = 1, \ldots, n$, where $g_{t_0}^t$ is a 2-parameter family of transformations in the evolution of the system and (e_i) is a standard basis in $\mathbb{R}^n$. In other words, we have the identities

$$\left|g^t_{t_0}(U)\right| = W(t) \cdot |U|, \quad \frac{\mathrm{d}}{\mathrm{d}t}\left|g^t_{t_0}(U)\right| = \mathrm{tr}A(t) \cdot \left|g^t_{t_0}(U)\right|, \qquad (6.28)$$

for a domain $U \subset \mathbb{R}^n$, where $|U|$ is the volume.

Let us apply this observation to the equation in variations with respect to initial conditions (6.13) in the case of the autonomous vector field $\dot{x} = v(x)$. This equation in variations has the form $\dot{y} = A(t)y$ (where $A(t) = \frac{\partial v}{\partial x}(\varphi_0(t))$ is the matrix of partial derivatives $\partial v_i / \partial x_j$ of the components v_i of the field) along the distinguished solution $\varphi_0(t)$. It is easy to check the identity

$$\mathrm{tr}A(t) = \sum_{i=1}^n \frac{\partial v_i}{\partial x_i}(\varphi_0(t)) = \mathrm{div}\, v(\varphi_0(t)), \qquad (6.29)$$

where div means the divergence.

Let $U \subset \mathbb{R}^n$ be a domain such that the solutions starting from U are defined for times between 0 and t. We divide the region U into rectangular cubes Δ_j, with small edge ε and with distinguished points $z_j \in \Delta_j$. Under the action of the flow g^t these cubes will become small nonlinear domains $g^t(\Delta_j)$, which are close to parallelepipeds spanned by the vectors $\varepsilon \cdot f_i(t)$, where each vector $f_i(t)$ is (as described above) associated with the equation in variations along the solution $\varphi_j(t)$ starting from z_j. Then we sum-up the volumes of the domains $g^t(\Delta_j)$ and go to the limit as $\varepsilon \to 0$, using the properties (6.28) and (6.29). As a result, we get the first part following statement (see Fig. 6.2).

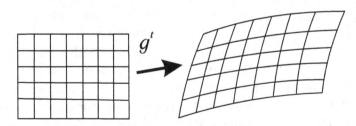

Fig. 6.2. Phase flow.

Theorem 6.24. *For the domain $U \subset \mathbb{R}^n$ and the flow g^t generated by the autonomous vector field $v(x)$ we have the identity*

$$\frac{\mathrm{d}}{\mathrm{d}t} \left| g^t(U) \right| = \int_{g^t(U)} \operatorname{div} v(x) \mathrm{d}^n x.$$

In particular, if $\operatorname{div} v(x) < 0$, then the flow g^t has the property of decreasing the volume, if $\operatorname{div} v(x) > 0$, then the flow increases the volume and, if $\operatorname{div} v \equiv 0$ then the Lebesque measure is invariant.

Moreover, if $\operatorname{div}(\rho v) \equiv 0$, where $\rho(x) \geq 0$ is a smooth function, then the measure $\rho(x) \mathrm{d}^n x$ is invariant with respect to the flow g_v^t.

To prove the second part we observe that the above Liouville formula amounts to

$$\mathcal{L}_v \mathrm{VOL} = \mathrm{d} \left(i_v \mathrm{VOL} \right),$$

where $\mathrm{VOL} = dx_1 \wedge \cdots \wedge dx_n$ is the volume form, $\mathcal{L}_X \eta = \frac{\mathrm{d}}{\mathrm{d}t} \big|_{t=0} \left(g_X^t \right)^* \eta$ is the Lie derivative of a k-form with respect to a vector field X and the $(k-1)$-form $(i_X \eta)_x (V_1, \ldots, V_{k-1}) = \eta_x (X(x), V_1, \ldots, V_{k-1})$ is the contraction of η by means of X.[c]

If ρ is a density, then we get

$$\mathcal{L}_v \left(\rho \cdot \mathrm{VOL} \right) = \mathrm{d} \left(\rho \cdot i_v \mathrm{VOL} \right) = \mathrm{d}\rho \wedge i_v \mathrm{VOL} + \rho \left(\mathrm{d} i_v \mathrm{VOL} \right)$$
$$= \left[v(\rho) + \rho \operatorname{div}(v) \right] \mathrm{VOL} = \operatorname{div}(\rho v) \cdot \mathrm{VOL}.$$

6.5. Lagrangian and Hamiltonian systems

6.5.1. *Newtonian, Lagrangian and Hamiltonian mechanics*

The simplest conservative Newtonian equation takes the form

$$m\ddot{x} = -U'(x),$$

where U is the potential. We know that it has the following first integral:

$$T + U = \frac{m}{2}\dot{x}^2 + U(x),$$

---

[c]It is a special case of the known homotopy formula $\mathcal{L}_X \eta = \mathrm{d}(i_X \eta) + i_X \mathrm{d}\eta$, whose proof uses the Stokes formula (see [11]).

where T denotes the kinetic energy. In the multi-dimensional case, e.g., for a mechanical system of many particles, we have the system of **Newtonian equations**

$$m_i \ddot{x}_i = -\partial U / \partial x_i, \quad i = 1, \ldots, n, \tag{6.30}$$

and the energy integral is again $T + U$, where $T = \sum \frac{1}{2} m_i \dot{x}^2$ and the potential $U = U(x_1, \ldots, x_n)$ is a function on the configuration space M of x's.

The well-known Hamilton principle of least action states that solutions to the system (6.30) are critical for certain **action functional** on the space of paths $\gamma : [t_0, t_1] \longmapsto M$ with fixed endpoints, $\gamma(t_0) = x_0$, $\gamma(t_1) = x_1$. This functional is the following integral:

$$S(\gamma) = \int_{t_0}^{t_1} L(\gamma(t), \dot{\gamma}(t)) \, \mathrm{d}t,$$

where

$$L = L(x, \dot{x}) = T - U$$

is the **Lagrange function** (or **Lagrangian**) of the mechanical system; here we admit that L can depend also on t. Recall also that the variational derivative of the functional S is the derivative in direction of a given vector,

$$S(\gamma + h) - S(\gamma) = \left\langle \frac{\delta S}{\delta \gamma}(\gamma), h \right\rangle + o(\|h\|).$$

Theorem 6.25. *If the variational derivative $\frac{\delta S}{\delta \gamma}(\gamma_0) = 0$, then the path γ_0 is a solution to the Newtonian equations*

In the proof, one writes

$$S(\gamma + h) - S(\gamma) = \int \left(\frac{\partial L}{\partial x}, h \right) \mathrm{d}t + \int \left(\frac{\partial L}{\partial \dot{x}}, h \right) \mathrm{d}t + o(\|h\|)$$

and rewrites the second integral above via the integration by parts formula:

$$\int \left(\frac{\partial L}{\partial \dot{x}}, h \right) \mathrm{d}t = \left(\frac{\partial L}{\partial \dot{x}}, h \right) \Big|_{t_0}^{t_1} - \int \left(\frac{\mathrm{d}}{\mathrm{d}t} \left(\frac{\partial L}{\partial \dot{x}} \right), h \right) \mathrm{d}t.$$

Since h is arbitrary such that $h(t_0) = h(t_1) = 0$ (and is small), the following **Euler–Lagrange equations**:

$$\frac{\mathrm{d}}{\mathrm{d}t}\left(\frac{\partial L}{\partial \dot{x}_j}\right) = \frac{\partial L}{\partial x_j}, \quad j = 1,\ldots,n \tag{6.31}$$

must hold for γ_0. It is clear that Eq. (6.31) are the same as Eq. (6.30).

The advantage of the latter Lagrangian approach over the Newtonian one is the possibility to work with general functions $q_1,\ldots,q_n$ on M as coordinates; they are called the **generalized coordinates**. Thus, we have $L = L(q,\dot{q})$ and system (6.31) is again the Euler–Lagrange system $\frac{\mathrm{d}}{\mathrm{d}t}(\partial L/\partial \dot{q}_j) = \partial L/\partial q_j$, $j = 1,\ldots,n$. With the generalized coordinates, the following **generalized momenta**

$$p_j = \frac{\partial L}{\partial \dot{q}_j}$$

are associated. For example, if $L = \sum \frac{1}{2}m_i\dot{q}_i^2 = U(q)$, then $p_i = m_i\dot{q}_i$.

Note also that if L does not depend on q_i (we say that the generalized coordinate q_i is cyclic) then the corresponding momentum p_i is a first integral (by the Euler–Lagrange equations).

When one rewrites T and U as functions of q_i and p_j then the total energy becomes a function of p and q:

$$H(p,q) = T + U. \tag{6.32}$$

The latter function is the so-called *Legendre transform of* L (with respect to $\dot{q}$):

$$H = \sup_{\dot{q}} \{(p,\dot{q}) - L(q,\dot{q})\};$$

here we use the fact that the kinetic energy is a convex function of $\dot{q}$. In the case of n particles with masses m_i and $q_i = x_i$, one gets

$$H = \sum \frac{p_i^2}{2m_i} + U(q).$$

Now, the Euler–Lagrange equations (6.31) become the **Hamiltonian equations**

$$\dot{q}_j = \frac{\partial H}{\partial p_j}, \quad \dot{p}_j = -\frac{\partial H}{\partial q_j}, \quad j = 1,\ldots,n. \tag{6.33}$$

Here $H = H(q,p,t)$ is the **Hamilton function** (or the **Hamiltonian**); if H does not depend directly on the time t, then system (6.33) is autonomous, otherwise, it is non-autonomous.

Indeed, we firstly solve the equation $p = (\partial L/\partial \dot{q})(q, \dot{q})$ with respect to $\dot{q}$, $\dot{q} = \dot{q}(p, q)$, and we get $H = (p, \dot{q}(p, q)) - L(q, \dot{q}(p, q)))$. Then we have $\partial H/\partial p = \dot{q} + (p - \partial L/\partial \dot{q})\partial \dot{q}/\partial p = \dot{q}$ and $\partial H/\partial q = -\partial L/\partial q + (p - \partial L/\partial \dot{q})\partial \dot{q}/\partial q = -\frac{d}{dt}(\partial L/\partial \dot{q}) = -\dot{p}$ (by the definition of p and the Euler–Lagrange equations).

6.5.2. Mechanics on Riemannian manifolds

Using the above approach, we can define a conservative mechanical system on any Riemannian manifold.

Recall that a **Riemannian manifold** M is equipped with the **Riemannian metric** which is a system of scalar products on the tangent spaces

$$(u, v)_q = g_q(u, v), \quad u, v \in T_q M.$$

If, in local coordinates $q_1, \ldots, q_n$ (associated with some chart on M) we write $u = \sum u_j \partial/\partial q_j$ and $v = \sum v_j \partial/\partial q_j$, then we have

$$g_q(u, v) = \sum g_{ij}(q) u_i v_j = (A(q)u, v), \tag{6.34}$$

where the matrix $A(q)$ is positive definite. It is used to write

$$ds^2 = \sum g_{ij}(q)\, dq_i dq_j,$$

where ds denotes an infinitesimal length.

The Riemannian metric allows to define the length of a curve γ in M joining two points $q^{(0)}$ and $q^{(1)}$:

$$L(\gamma) = \int_{t_0}^{t_1} \sqrt{|\dot{\gamma}(t)|^2_{\gamma(t)}}\, dt. \tag{6.35}$$

It also allows to define the kinetic energy of a material particle (of mass $m = 1$) moving on M:

$$T(q, \dot{q}) = \frac{1}{2}(A(q)\dot{q}, \dot{q}). \tag{6.36}$$

A conservative mechanical system on M is a Lagrangian system defined by the Lagrange function $L = T - U$, where $U = U(q)$ is the potential function on M. Here the Lagrangian is treated as a function on the tangent bundle to $M : L : TM \longmapsto \mathbb{R}$.

The generalized momenta are defined as before, $p_j = \partial L/\partial \dot{q}_j$, but now they are elements of the cotangent space $T_q^* M$ (as linear functionals). We have

$$p = A(q)\dot{q}, \quad \dot{q} = A^{-1}(q)p, \quad T = \frac{1}{2}\left(A^{-1}(q)p, p\right).$$

The Hamilton function $H(p,q) = T(p,q) + U(q)$ is a function on the cotangent bundle: $H : T^* M \longmapsto \mathbb{R}$.

In the case $U \equiv 0$, the above action functional takes the form $S(\gamma) = \int T(\gamma, \dot{\gamma})\, dt$. It turns out that the corresponding Euler–Lagrange equations (i.e., with $L = T$) are the same as the Euler–Lagrange equations for the functional (6.35) (i.e., with $L = \sqrt{2T}$), because the function $2T$ (twice the total energy) is constant.

In the case $L = T$, the Euler–Lagrange equations are the equations for the geodesic lines,

$$\frac{\mathrm{d}}{\mathrm{d}t}\left(A(q)\dot{q}\right) = \left(\frac{\partial A}{\partial q}\dot{q}, \dot{q}\right)$$

and the corresponding second-order differential equations are expressed via the Christoffel symbols. The corresponding flow $\{g^t\}$ on the energy surface $\{T = 1/2\}$ (unit tangent bundle which is invariant) is called the **geodesic flow**. We have $g^t(q,v) = (q(t), v(t))$, where $q(t)$ moves with velocity 1 along the geodesic line staring at q in the direction of v and $v(t)$ is the unit vector at $q(t)$ tangent to this geodesic line.

The corresponding Hamilton function equals $H(p,q) = \frac{1}{2}\left(A^{-1}(q)p, p\right)$ and the corresponding Hamiltonian equations

$$\dot{q} = A^{-1}(q)p, \quad \dot{p} = -\frac{1}{2}\left(\frac{\partial A^{-1}}{\partial q}p, p\right)$$

define the **geodesic flow** on the unit cotangent bundle $\{(q,\omega) \in T^* M : |\omega| = 1\}$.

6.5.3. *Canonical formalism*

Important advantage of using Hamiltonian systems, in contrast to Newtonian ones (or, more generally, the so-called Euler–Lagrange systems), is that one can simplify (or integrate) them using changes

$$g : (Q, P) \longmapsto (q, p), \tag{6.37}$$

$q = q(Q, P)$, $p = p(Q, P)$, which can mix generalized coordinates and generalized momenta.

Definition 6.26. The change g is **symplectic** (or canonical) if

$$\sum \mathrm{d}p_j \wedge \mathrm{d}q_j = \sum \mathrm{d}P_j \wedge \mathrm{d}Q_j,$$

i.e., if $g^*\omega = \omega$ for the differential 2-form

$$\omega = \sum \mathrm{d}p_j \wedge \mathrm{d}q_j = \mathrm{d}p \wedge \mathrm{d}q. \tag{6.38}$$

Theorem 6.27. *If the change* (6.37) *is symplectic and the variables* q_j, p_j *satisfy the Hamiltonian system with a Hamilton function* H, *then the variables* Q_j, P_j *satisfy the Hamilton system with the Hamilton function* $K(Q, P, t) = H(q, p, t)$, *i.e.,* $K = g^*H$.

In the proof, one considers the differential 1-form

$$\eta = p\mathrm{d}q - H\mathrm{d}t = \sum p_j \mathrm{d}q_j - H\mathrm{d}t \tag{6.39}$$

in $\mathbb{R}^{2n+1}$. Its external derivative $\mathrm{d}\eta = \omega - \mathrm{d}H \wedge \mathrm{d}t$, at each point (q, p, t) has one-dimensional kernel spanned by the vector $v = \left(H'_p, -H'_q, 1\right)$, i.e., $\langle \mathrm{d}\eta, v, w \rangle = 0$ for any vector w (Exercise 6.62). The same takes place with the form $g^*\eta = P\mathrm{d}Q - K\mathrm{d}t$. See [11] for more details.

6.5.4. *Poisson brackets and integrability*

Assume that the function $H = H(q, p)$ does not depend directly on t. It is easy that in this case this function is a first integral for the corresponding Hamiltonian vector field

$$X_H = H'_p \partial/\partial q - H'_q \partial/\partial p,$$

i.e., $\partial H/\partial X_H = X_H H = 0$ (Exercise 6.59). We have a special notion for such derivative.

Definition 6.28. The **Poisson bracket** of two functions F and G on $\mathbb{R}^{2n}$ is defined as

$$\{F, G\} := X_G F = \sum \frac{\partial F}{\partial q_j} \frac{\partial G}{\partial p_j} - \frac{\partial F}{\partial p_j} \frac{\partial G}{\partial q_j}. \tag{6.40}$$

If $\{F, G\} = 0$, then we say that these functions are **in involution**.

The Poisson bracket is antisymmetric, which is one of the conditions for the Lie algebra structure on the space $C^\infty \left(\mathbb{R}^{2n}\right)$ with the commutator defined by it. It turns out that it satisfies other conditions.

Theorem 6.29. *The above Poisson bracket has the following properties*:

$$\{FG, H\} = F\{G, H\} + G\{F, H\} \tag{6.41}$$

and

$$\{\{F, G\}, H\} + \{\{G, H\}, F\} + \{\{H, F\}, G\} = 0, \tag{6.42}$$

i.e., the **Jacobi identity**.

The property (6.41) is obvious. To prove the second identity, we will look more closely at the left-hand side. It is a linear combination of second-order derivatives of the three functions (with combinations of products of first-order derivatives of the other functions as coefficients). Let us look for the part containing the second-order derivatives of F:

$$\{\{F, G\}, H\} - \{\{F, H\}, G\} = (X_H X_G - X_G X_H) F.$$

It is the commutator $[X_G, X_H]$ of first-order differential operators acting on the function F. But, in fact, such commutator is a first-order differential operator, because the second-order derivatives cancel themselves (Exercise 6.63).

The Jacobi identity implies the following formula for the commutator of two Hamiltonian vector fields:

$$[X_F, X_G] = X_{\{G, F\}},$$

i.e., it is again a Hamiltonian vector field with the minus Poisson bracket as the Hamilton function. Indeed,

$$X_{\{G, F\}} H = \{H, \{G, F\}\} = \{\{H, G\}, F\} - \{\{H, F\}, G\}$$
$$= X_F X_G H - X_G X_F H = [X_F, X_G] H.$$

Next, we note the following theorem.

Theorem 6.30. *If the commutator of two vector fields X and Y vanishes, then their phase flows commute:*

$$g_X^t \circ g_Y^s = g_Y^s \circ g_X^t.$$

In the proof, one divides the rectangle $[0, t] \times [0, s]$ into squares with sides of small length ε; there are $O(\varepsilon^{-2})$ of them. Next, one performs a transition from the two sides $\{0\} \times [0, s] \cup [0, t] \times \{s\}$ of the rectangle (corresponding to the left-hand side of the above equality) to the opposite two sides (corresponding to the right-hand side) by a series of changes, each relying upon analogous changes of sides of the small squares; there are $O(\varepsilon^{-2})$ such changes. The property $[X, Y] = 0$ implies that the quantity $g_X^\varepsilon \circ g_Y^\varepsilon \circ g_X^{-\varepsilon} \circ g_Y^{-\varepsilon} - id$ is of order $O(\varepsilon^3)$.

In the following, we repeat some notions from the beginning of Chapter 4.

Definition 6.31. An autonomous Hamiltonian vector field X_H is called **completely integrable** if there exists a collection $F_1 = H, \ldots, F_n$, such that:

- they are in involution, $\{F_i, F_j\} = 0$;
- they are independent, i.e., their differentials $dF_i(q, p)$ are independent for generic point (q, p).

Theorem 6.32 (Liouville–Arnold). *If the common levels $\{F_1 = c_1, \ldots, F_n = c_n\}$ of a completely integrable Hamiltonian system are connected, compact and smooth, then they are tori $\mathbb{T}^n$.*

*Moreover, in a neighborhood of a given torus there exists a system of coordinates $(I_1, \ldots, I_n, \varphi_1, \ldots, \varphi_n)$, the so-called **action–angle variables**, in which the initial Hamiltonian system takes the following Hamiltonian form:*

$$\dot{I}_j = 0, \ \dot{\varphi}_j = \omega_j(I) = \partial H_0/\partial I_j, \ j = 1, \ldots, n,$$

where $H_0(I_1, \ldots, I_n) = H(q, p)$ is the Hamiltonian after the change. In general, the movement on the tori $\{I_1 = d_1, \ldots, I_n = d_n\}$, which are parameterized by the angles φ_j mod 2π, is periodic or quasi-periodic:

$$\varphi_j(t) = \varphi_j(0) + \omega_j(I)t.$$

In the proof, one uses the fact that the phase flows $\{g_j^t\}$ of the vector fields X_{F_j} commute (by Theorems 6.29 and 6.30). Consider a common level $M_c = \{F_1 = c_1, \ldots, F_n = c_n\}$, like in the theorem, and a point $a \in M_c$. We define the map $G : \mathbb{R}^n \longmapsto M_c$,

$$(t_1, \ldots, t_n) \longmapsto g_n^{t_n} \circ \cdots \circ g_1^{t_1}(a).$$

It turns out to be surjective and a local diffeomorphism, hence a topological covering. So, we have $M_c \simeq \mathbb{R}^n/\text{stab}(a)$, the quotient of the Euclidean space by the stabilizer of the point a. The latter stabilizer is discrete and hence forms a lattice $\mathbb{Z}v_1 + \cdots + \mathbb{Z}v_n$, where v_j are independent vectors. So, $\mathbb{R}^n/\text{stab}(a)$ is a torus.

The constant vector field $\partial/\partial t_1$ (in the coordinates t_j) becomes a constant vector field $\sum \omega_j \partial/\partial \varphi_j$ in an angular coordinate system associated with the basis $\{v_j\}$: $(t_1, \ldots, t_n) = \sum (\varphi_j/2\pi)v_j$ modulo $\text{stab}(a)$.

For the construction of the action–angle variables, we refer the reader to [6].

6.5.5.   *Symplectic and Poisson structures*

The notion of a Hamiltonian system is generalized as follows.

Definition 6.33. A **symplectic structure** on a differentiable manifold N is a differential 2-form ω (**symplectic form**) with the following properties:

(i) it is closed, $d\omega = 0$;
(ii) it is *non-degenerate* in the sense that the following linear map:

$$v \longmapsto i_v\omega = \omega(v, \cdot)$$

from $T_x N$ to $T_x^* N$ is an isomorphism for any $x \in N$.

Then (N, ω) is a **symplectic manifold**.

A natural example of a symplectic manifold is the cotangent bundle to a manifold, $N = T^*M$. We have the following *Liouville 1-form* η such that $\eta(v) = p(D\pi \cdot v)$; here $\pi : T^*M \longmapsto M$ is the fiber bundle projection, $v \in T_{(x,p)}N$ is a tangent vector at the point $p \in T_x^*M$

and $D\pi$ is the derivative of the projection. The form

$$\omega = \mathrm{d}\eta$$

is the symplectic form on T^*M. In local coordinates, $q_1, \ldots, q_n$ in the base M and $p = p_1 \mathrm{d}q_1 + \cdots + p_n \mathrm{d}q_n$ in the fiber T_q^*M (where $\{\mathrm{d}q_j\}$ is the dual basis to the basis $\{\partial/\partial q_j\}$ in $T_q N$) we have $\eta = p_1 \mathrm{d}q_1 + \cdots + p_n \mathrm{d}q_n = p \mathrm{d}q$ and hence $\omega = \mathrm{d}p_1 \wedge \mathrm{d}q_1 + \cdots + \mathrm{d}p_n \wedge \mathrm{d}q_n = \mathrm{d}p \wedge \mathrm{d}q$ (like in Eq. (6.38)).

This is a special case of the following more general result.

Theorem 6.34 (Darboux). *If ω is a symplectic form on N, then for any point $x_0 \in N$ there exists a diffeomorphism $h : U \longmapsto V \subset \mathbb{R}^{2n}$, $2n = \dim N$, from a neighborhood U of x_0 to a neighborhood of 0 in $\mathbb{R}^{2n}$ (with coordinates p, q) such that*

$$h^* \mathrm{d}p \wedge \mathrm{d}q = \omega.$$

In the proof, we can assume that we have a non-degenerate and closed 2-form ω in a neighborhood of 0 in $\mathbb{R}^{2n}$. Let ω_0 be a constant 2-form such that $\omega_0 = \omega(0)$ and let $\omega_1 = \omega - \omega_0$. Consider the 1-parameter family of closed differential forms $\omega_t = \omega_0 + t\omega_1$, $0 \leq t \leq 1$. We look for a family $\{f_t\}$ of diffeomorphisms such that

$$f_t^* \omega_0 = \omega_t$$

and $f_t(0) = 0$. Then $h = f_1^{-1}$ transforms ω to the constant form ω_0 which can be reduced to $\mathrm{d}p \wedge \mathrm{d}q$ by a construction of the corresponding symplectic basis (Exercise 6.64).

Let $v_t(x) = \frac{\mathrm{d}}{\mathrm{d}t} f_t(x)$ be the corresponding (non-autonomous in general) vector field and let α be a 1-form such that $\omega_1 = \mathrm{d}\alpha$, by the Poincaré lemma, and $\alpha(0) = 0$ (here the closeness of ω is used). Differentiating the both sides of the above equation with respect to t we find $\mathcal{L}_{v_t}\omega_0 = \omega_1$, where $\mathcal{L}$ denoted the Lie derivative, and using the homotopy formula with $\mathrm{d}\omega_0 = 0$ (see Note above), we get $\mathrm{d}i_{v_t}\omega_0 = \mathrm{d}\alpha$. By the non-degeneracy of ω_0 the equation $i_{v_t}\omega_0 = \alpha$ is solvable with respect to the vector field v_t such that $v_t(0) = 0$. In fact, it is an autonomous vector field, $v_t = v$, and generates the flow $f_t = g_v^t$ such that $g_v^t(0) = 0$.

For a symplectic form we define the **Hamiltonian vector field** X_H by

$$i_{X_H}\omega = -\mathrm{d}H \qquad (6.43)$$

and the **Poisson bracket** by

$$\{F, G\} = X_G F. \qquad (6.44)$$

The latter Poisson bracket satisfies conditions (6.41)–(6.42) from Theorem 6.29. Therefore, other results, e.g., about invariant tori and action–angle coordinates (the Liouville–Arnold theorem) hold true.

But we have a further generalization of the above theory.

Definition 6.35. A **Poisson structure** on a differentiable manifold N is a map $\{\cdot, \cdot\} : C^\infty(N) \times C^\infty(N) \longmapsto C^\infty(N)$ (the **Poisson bracket**) with the following properties:

(i) it is bilinear and antisymmetric, $\{F, G\} = -\{G, F\}$;
(ii) it satisfies the Leibnitz identity, $\{FG, H\} = F\{G, H\} + G\{F, H\}$;
(iii) it satisfies the Jacobi identity, $\{F, \{G, H\}\} = \{\{F, G\}, H\} + \{G, \{F, H\}\}$ ($\{F, \cdot\}$ is a differentiation).

It is quite clear, knowing the above theorems, that Eq. (6.44) defines a bracket which obeys conditions from the latter definition. In the symplectic case with the standard symplectic form $\mathrm{d}p \wedge \mathrm{d}q$, we have the following brackets (Exercise 6.65):

$$\{p_i, q_j\} = \delta_{ij}, \ \{p_i, p_j\} = \{q_i, q_j\} = 0. \qquad (6.45)$$

The Poisson bracket allows to define the Hamiltonian vector field X_H by Eq. (6.44), i.e.,

$$\dot{F} = X_H F = \{F, H\}$$

for any function F on N (e.g., a coordinate); here H is the Hamiltonian function.

But not every Poisson bracket obeying the properties (i)–(iii) above arises form some symplectic structure, because there is no any version of the non-degeneracy condition from Definition 6.33.

In particular, it is possible that some functions have zero Poisson with any other functions on N.

Definition 6.36. A function G on a Poisson manifold which has zero Poisson bracket with any other function is called a **Casimir function**. The Casimir functions form an algebra generated by some finite collection of Casimir functions $G_1, \ldots, G_k$. The common levels $\{G_1 = c_1, \ldots, G_k = c_k\}$ of the Casimir functions are called **symplectic leaves**.

Proposition 6.37. *The Poisson bracket restricted to functions on a symplectic leaf arises from a non-degenerate symplectic structure on this leaf.*

6.6. Exercises

Exercise 6.38. Depending on the constants $M = \sup_{I \times U} |v(t, x)|$ and $L = \sup |v(t, x_1) - v(t, x_2)| / |x_1 - x_2|$ (the Lipschitz constant), choose ε in $I_0 = (t_0 - \varepsilon, t_0 + \varepsilon)$ and the radii of balls $U_0 = B(x_*, r) = \{|x - x_*| < r\} \subset U$ and $\mathcal{B}(x_0, R) = \{\varphi : I_0 \times U_0 \longmapsto \mathbb{R}^n : \sup |\varphi(t, x_0) - x_0| < R\}$ such that:

(1) $\mathcal{T} : \mathcal{B}(x_0, R) \longmapsto \mathcal{B}(x_0, R)$ and
(2) $\mathcal{T}$ is a contraction on $\mathcal{B}(x_0, R)$.

This will complete the proof of Theorem 6.5.

Exercise 6.39. Complete the proofs of Theorems 6.7 and 6.8.
Hint: In the proof of Theorem 6.7, consider the sequences $x = \varphi_n(t; x_0)$, $z = \psi_n(t; x_0) = \partial \varphi_n / \partial x_0$ of approximations to the solutions to the initial problem $\dot{x} = v(t; x)$, $\dot{z} = \frac{\partial v}{\partial x}(t, x)z$, $x(t_0) = x_0$, $z(t_0) = I$, where $z(t; x_0)$ takes values in the spaces of the matrices $n \times n$. In the proof of Theorem 6.8, use Theorem 6.7.

Exercise 6.40. Prove that, if $v(t, x; \lambda)$ depends analytically on the arguments, then the solution $\varphi(t; x_0; \lambda)$ is also analytic.

Exercise 6.41. Complete the proof of Theorem 6.9.

Exercise 6.42. Complete the proof of Theorem 6.10.

Exercise 6.43. Find the solution of the equation $x^2 \frac{dy}{dx} - \cos 2y = 1$ satisfying the condition $y(+\infty) = \frac{9\pi}{4}$.

Exercise 6.44. Solve the equation $\frac{dy}{dx} = \sqrt{4x + 2y - 1}$.

Exercise 6.45. Solve the equation $x\frac{dy}{dx} = y - xe^{y/x}$.

Exercise 6.46. Solve the equation $\frac{dy}{dx} = y^2 - 2/x^2$.

Exercise 6.47. Solve the equation $2ydx + (x^2y + 1)xdy = 0$.

Exercise 6.48. Solve the equation $xy' - 2y = 2x^4$.

Exercise 6.49. Solve the equation $xydy = (y^2 + x)dx$.

Exercise 6.50. Solve the Riccati equation $y' = 2xy - y^2 + 5 - x^2$. *Hint*: Guess one solution.

Exercise 6.51. Solve the equation $dy/dx = \left(ax^2 + by^2 + 1\right)/2xy$. *Hint*: Look for an integral factor in the form of x^α.

Exercise 6.52. Solve the equation $\frac{y}{x}dx + (y^3 + \ln x)dy = 0$.

Exercise 6.53. Consider a linear system $\dot{x} = A(t)x + b(t)$, with continuous $A(t)$ and $b(t)$ and with estimates $\|A(t)\| \leq C_1(t)$ and $|b(t)| \leq C_2(t)$. Show the estimates $\left|\frac{d}{dt}|x|^2\right| \leq 2C_1(t)|x|^2 + 2C_2(t)|x| \leq C_3(t)|x|^2$, where the last inequality holds for sufficiently large $|x|$ and a certain continuous function $C_3(t)$. Deduce from this that solutions cannot escape to infinity after a finite time.

Exercise 6.54. Give a general solution of the system $\dot{x} = x - y - z$, $\dot{y} = x + y$, $\dot{z} = 3x + z$.

Exercise 6.55. Give a general solution of the system $\dot{x} = x - y + 1/\cos t$, $\dot{y} = 2x - y$.

Exercise 6.56. Give a general solution of the equation $d^4x/dt^4 + 4x = 0$.

Exercise 6.57. Give a general solution of the equation $\ddot{x} + 2\dot{x} + x = t(e^{-t} - \cos t)$.

Exercise 6.58. For which k and ω the equation $\ddot{x} + k^2x = \sin\omega t$ has at least one periodic solution?

Exercise 6.59. Show that, if the Hamilton function does not depend directly on time, then it is the first integral for the corresponding Hamiltonian system.

Exercise 6.60. Show that any Hamiltonian vector field has zero divergence. Deduce from this that the corresponding phase flow preserves the $2n$-dimensional volume.

Exercise 6.61. Show that, if H does not depend directly on time, then the equilibrium points of system (6.33) are exactly the critical points of the function H.

Exercise 6.62. Prove that the kernel of the form η from Eq. (6.39) is one-dimensional and is generated by the vector $\left(H'_q, -H'_p, 1\right)$.

Exercise 6.63. Give a formula for the commutator $[Y, Z]$ of two vector fields $Y = \sum Y_j(x)\partial/\partial x_j$ and $Z = \sum Z_j(x)\partial/\partial x_j$.

Exercise 6.64. Let ω be a non-degenerate 2-form with constant coefficients. Show that there exists a system $(e_1, f_1, \ldots, e_n, f_n)$ such that $\omega(e_i, e_j) = \omega(f_i, f_j) = 0$, $\omega(e_i, f_j) = \delta_{ij}$. In particular, the dimension of a symplectic manifold is even.
Hint: Use induction with respect to the dimension.

Exercise 6.65. Prove formulas (6.45).

References

[1] V. I. Arnold and Y. S. Ilyashenko, *Dynamical Systems I: Ordinary Differential Equations and Smooth Dynamical Systems, Encyclopaedia of Mathematical Sciences*, Vol. 1. Springer Science & Business Media, New York (1988), translated from Russian.

[2] A. F. Filippov, *Recueil de Problèmes d'Équations Differéntielles*. Mir, Moscow (1976), translated from Russian.

[3] L. S. Pontryagin, *Ordinary Differential Equations*. Addison-Wesley, C.A., Palo Alto (1962), translated from Russian.

[4] W. Szlenk, *An Introduction to the Theory of Smooth Dynamical Systems*. John Wiley & Sons, New York (1984), translated from Polish.

[5] P. Hartman, *Ordinary Differential Equations*. John Wiley & Sons, New York (1964).

[6] V. I. Arnold, *Mathematical Methods of Classical Mechanics*, Vol. 60. Springer Science & Business Media, New York (1989), translated from Russian.

[7] D. K. Arrowsmith and C. M. Place, *Ordinary Differential Equations. A Qualitative Approach with Applications*. Chapman & Hall, London (1982).

[8] L. Perko, *Differential Equations and Dynamical Systems*. Springer Science & Business Media, New York (2001).

[9] A. A. Andronov, E. A. Leontovich, I. I. Gordon and A. G. Maier, *Qualitative Theory of Second-Order Dynamic Systems*. Halsted Press, New York (1973), translated from Russian.

[10] A. A. Andronov, E. A. Leontovich, I. I. Gordon and A. G. Maier, *Theory of Bifurcations of Dynamical Systems on a Plane*. John Wiley & Sons, New York (1973), translated from Russian.

[11] V. I. Arnold, *Geometrical Methods in the Theory of Ordinary Differential Equations*, Vol. 250. Springer Science & Business Media, New York (1983), translated from Russian.

[12] J. E. Marsden and M. McCracken, *The Hopf Bifurcation and its Applications (With contributions by P. Chernoff, G. Childs, S. Chow, J. R. Dorroh, J. Guckenheimer, L. Howard, N. Kopell, O. Lanford, J. Mallet-Paret, G. Oster, O. Ruiz, S. Schecter, D. Schmidt and S. Smale)*, Vol. 19. Springer Science & Business Media (1976).

[13] N. Bautin and E. Leontovich, *Metody i Priemy Kachestvennogo Issledovaniya Dinamicheskikh Sistem na Ploskosti (Methods and Techniques of the Qualitative Study of Dynamical Systems on the Plane), Spravochnaya Matematicheskaya Biblioteka (Mathematical Reference Library)*, Vol. 11. Nauka, Moscow (1990), Russian.

[14] J. Guckenheimer and P. Holmes, *Nonlinear Oscillations, Dynamical Systems, and Bifurcations of Vector Fields*, Vol. 42. Springer Science & Business Media, New York (2013).

[15] V. I. Arnold, V. S. Afrajmovich, Y. S. Ilyashenko and L. P. Shilnikov, *Dynamical Systems V: Bifurcation Theory and Catastrophe Theory, Encyclopaedia of Mathematical Sciences*, Vol. 5. Springer Science & Business Media, New York (1994), translated from Russian.

[16] R. L. Devaney, *An Introduction to Chaotic Dynamical System.* Benjamin/Cummings Co. Inc., Menlo Park, CA (1986).

[17] C. Robinson, *Dynamical Systems. Stability, Symbolic Dynamics and Chaos.* CRC Press, Boca Raton (1998).

[18] I. P. Cornfeld, S. V. Fomin and Y. G. Sinai, *Ergodic Theory*, Vol. 245. Springer Science & Business Media, New York (1982), translated from Russian.

[19] Z. Nitecki, *Differentiable Dynamics. An Introduction to the Orbit Structure of Diffeomorphisms.* The MIT Press, New York (1971).

[20] F. Brauer and J. A. Nohel, *The Qualitative Theory of Ordinary Differential Equations: An Introduction.* Courier Corporation, New York (1989).

[21] F. Dumortier, J. Llibre and J. C. Artés, *Qualitative Theory of Planar Differential Systems.* Springer Science & Business Media, New York (2006).

[22] J. K. Hale, *Ordinary Differential Equations.* Robert E. Krieger Publishing Co. Inc., New York (1980).

Index

A

Abelian integral, 48, 119, 126–127,
150, 156
absolute, 224
action functional, 250
action–angle variables, 134, 137, 144,
257
angular momentum, 136, 139, 141
Anosov flow, 221
attractor
hyperbolic, 229
Lorenz, 230
strange, 229

B

bifurcation
Andronov–Hopf, 100, 108
subcritical, 104
supercritical, 105
Bogdanov–Takens, 117, 156
codimension 2, 117
Hopf, 107
Hopf–Hopf, 123
local, 100
non-local, 101, 112
period doubling, 100, 113
pitchfork, 114
saddle-node, 100–101
transcritical, 102
saddle-node for limit cycle, 100,
112
saddle-node–Hopf, 120
blowing-up of singularity, 68

C

characteristic exponent, 58
Chebyshev space, 156
commutator of vector fields,
255
conjugacy, 26
critical point, 87
critical value, 87
cyclic coordinate, 136
cyclicity, 41
cylindrical set, 210

D

diffeomorphism, 99
Anosov, 221
Axiom A, 220
Morse–Smale, 221
duck cycle, 186
dynamical system
ergodic, 193
metric, 192
mixing, 216
symbolic, 209, 217
transitive, 193, 211

E

elliptic integral, 151
entropy
  metric, 217
  topological, 218
equation
  Bernoulli, 242
  in variations, 238
  Lienard, 46
  linear, 241, 244
  Newtonian, 250
  normal variations, 169
  Pfaffian, 149, 240
  quasi-homogeneous, 241
  Riccati, 176, 182
  van der Pol, 62
equilibrium point, 33
  asymptotically stable, 2
  Bogdanov–Takens, 77
  center, 35, 66–67
  degenerate, 66
  elementary, 66
  focus, 66
    strong, 40
    weak, 40
  hyperbolic, 17
  Lyapunov stable, 2
  monodromic, 71
  node, 66, 97
  saddle, 66, 99, 131
  saddle-node, 66, 99, 130
ergodic mean, 194
Euler case, 141
even distribution, 194

F

fast variable, 179
fixed point
  hyperbolic, 19
function
  Casimir, 141, 260
  Darboux, 45, 172
  Dulac, 61
  Hamilton, 134, 243, 251
  Lagrange, 145, 250

Lyapunov, 4
Melnikov, 48, 121, 124, 150, 157,
  169, 204
Weierstrass, 143, 176
functions in involution, 254
fundamental matrix, 170, 245

G

Gelfand–Leray form, 153, 212
generalized coordinate, 251
generalized momentum, 251
geodesic flow, 224, 253

H

harmonic oscillator, 2, 135
Hénon map, 230
Hess surface, 173
Hess–Appelrot case, 173
Hilbert problem, 48, 150
homological operator, 98, 118, 120,
  123, 130
horocycle, 226
Hurwitz matrix, 10
hyperbolic set, 219
hyperbolic torus automorphism,
  221

I

index, 56
induced family, 80
inertia tensor, 139–140
integrating factor, 242
invariant manifold
  center, 92
  stable, 17
  unstable, 17

J

Jacobi identity, 140, 255, 259
jet, 90

K

Kepler problem, 135
Kovalevskaya case, 141

L

Lagrange case, 141, 164
Legendre transform, 251
libration point
  collinear, 161
  triangular, 161
limit cycle, 35
limit set, 49
Lobachevskian plane, 224
loss of stability, 183
Lyapunov exponent, 168

M

Markov chain, 213
Markov partition, 220
measure
  Bernoulli, 213
  invariant, 193, 211
  Liouville, 249
monodromy map, 157, 170, 177, 201

N

non-wandering set, 197, 220
normal form
  Birkhoff, 162
  Poincaré–Dulac, 96, 104
normal hyperbolicity, 167

O

orbit, 26
orbital equivalence, 73, 80

P

pendulum, 33
period, 33
periodic point
  hyperbolic, 19
periodic solution, 33
phase curve
  hyperbolic, 37
phase flow, 235
phase portrait, 33
Poincaré plane, 73
Poincaré return map, 36

Poincaré–Lyapunov focus quantity, 40
Poisson bracket, 134, 140, 254, 259
Poisson structure, 259

R

relaxation oscillations, 179
resolution of singularity, 68
resonance, 95, 163
  strong, 101, 125
  weak, 100, 114
Riemannian manifold, 252
Riemannian metric, 252
rigid body, 138, 173
rotation number, 195

S

sector
  elliptic, 71
  hyperbolic, 71
  parabolic, 71
selenoid, 229
separatrix, 71
separatrix connection, 72, 101, 115
separatrix loop, 72, 101, 115
slow surface, 180
slow variable, 179
Smale horseshoe, 208
spurt, 182
stability
  orbital structural, 73
  structural, 27
subharmonic orbit, 156
swing, 201
symplectic form, 257
symplectic leaf, 141, 260
symplectic map, 163, 254
symplectic structure, 257
system
  averaged, 147
  Duffing, 75, 201
  Euler–Lagrange, 251
  Euler–Poisson, 140–141
  FitzHugh–Nagumo, 54, 108
  fundamental, 244

Hamiltonian, 133, 243, 251
  completely integrable, 134, 256
Hess–Appelrot, 175
Jouanolou, 60
Lagrange, 143
Lienard, 46, 51
linear, 244
Lotka–Volterra, 30, 45, 61, 171
partially integrable, 167
slow–fast, 179
van der Pol, 133, 135, 179
Zhukovskii, 109

T

Theorem
  about averaging, 147
  about phase flow, 239
  about prolongation, 240
  Andronov–Hopf, 104
  Anosov, 222
  Arnold–Liouville, 146
  Birkhoff, 163
  Birkhoff ergodic, 215
  Bogolyubov–Krylov, 212
  Bowen, 220
  Chetaev, 7
  Darboux, 258
  Denjoy, 198
  Dulac, 58
  ergodic for Markov chains, 214
  flow box, 239
  Grobman–Hartman, 23
  Hadamard–Perron, 18
  Hirsch–Pugh–Shub, 168

Kolmogorov–Arnold–Moser, 159,
  162, 165
Liouville, 247
Liouville–Arnold, 134
Lyapunov, 4
Lyapunov–Schmidt, 93
  on dependence on initial
    conditions, 237
  on dependence on parameters,
    238
  on existence and uniqueness,
    237
Poincaré recurrence, 214
Poincaré–Bendixson, 49
Poincaré–Dulac, 96
Poincaré–Lyapunov, 40
Robbin, 221
Routh–Hurwitz, 10
Sard, 87
Shoshitaishvili, 92
Singer, 244
Takens, 117
Thom, 85, 90
three body problem, 160
transversality, 84, 90

V

vector field
  Hamiltonian, 140, 259
versal family, 81

W

wandering interval, 197
Wronskian, 245

Printed in the United States
by Baker & Taylor Publisher Services